FOUCAULT, BIOPOLITICS, AND
GOVERNMENTALITY

———

SÖDERTÖRN
PHILOSOPHICAL STUDIES
2013

Foucault, Biopolitics, and Governmentality

Edited by
Jakob Nilsson &
Sven-Olov Wallenstein

SÖDERTÖRN
PHILOSOPHICAL STUDIES
14

Other titles in this series

Rethinking Time: Essays on
History, Memory, and Representation (2011)
Hans Ruin & Andrus Ers (eds.)

Phenomenology of Eros (2012)
Jonna Bornemark & Marcia Sá Cavalcante Schuback (eds.)

Ambiguity of the Sacred (2012)
Jonna Bornemark & Hans Ruin (eds.)

Translating Hegel (2013)
Brian Manning Delaney & Sven-Olov Wallenstein (eds.)

———————————————

Södertörn University
The Library
SE-141 89 Huddinge

www.sh.se/publications

Cover image: Cecilia Grönberg and Jonas (J) Magnusson,
from the series Telefonfabriken: Produktion/postproduktion: 1938–2003,
Graphic Form: Jonathan Robson & Per Lindblom
Printed by E-print, Stockholm 2013

Södertörn Philosophical Studies 14
ISSN 1651-6834

Södertörn Academic Studies 54
ISSN 1650-433X

ISBN 978-91-86069-59-9 (print)
ISBN 978-91-86069-61-2 (digtal)

Contents

Introduction: Foucault, Biopolitics, and Governmentality

Sven-Olov Wallenstein

The idea of biopolitics

Foucault's analysis of biopolitics has produced a huge amount of responses, and similar to his closely connected discussions of governmentality, it has engendered a body of literature, even entire schools of thought that have evolved far beyond the limits of his own proposals.[1] It is true that, even before Foucault enters the scene, biopolitics has its own history; in many respects, then, Foucault's work constitutes an intervention into an already complex tradition.[2] His contributions have nevertheless become a focal

---

[1] For overviews, see Renata Brandimarte et al (eds.), *Lessico di Biopolitica* (Rome: Manifestolibri, 2006), Antonella Cutro (ed.), *Biopolitica: Storia e attualità di un concetto* (Verona: Ombrecorte, 2005), and Thomas Lemke, "From State Biology to the Government of Life: Historical Dimensions and Contemporary Perspectives of 'Biopolitics'," *Journal of Classical Sociology*, Vol. 10, No. 4 (2010). A rich material can also be found in the two series of periodical publications that have been dedicated to the topic: *Zeitschrift für Biopolitik*, edited by Andreas Mietzsch, and *Research in Biopolitics*, edited by Albert Somit.

[2] The term "biopolitics" seems to have been invented by the Swedish political scientist Rudolf Kjellén (1864–1922), who understands it on the basis of an organicist conception of the state as "life-form" and "ethnic individuality." This life-form comprises social groups that struggle for existence, but also cooperate, in a dynamic process that makes up the life of state, which Kjellén proposes to study under the rubric "biopolitics." Kjellén's main work on the theme is *Staten som lifsform* (Stockholm: Hugo Geber, 1916); on Kjellén, see Fredrika Lagergren, *The Versatile Scientist: Rudolf Kjellén and the Theory of Biopolitics* (Stockholm: Kungl. Vitterhets-, historie- och antikvitetsakademien, 1998). Similar conceptions were widespread at the time, and political theories were also proposed by professional biologists, such as Jakob von Uexküll's *Staatsbiologie (Anatomie, Physiologie, Pathologie des Staates)* (Berlin: Gebr. Paetel, 1920), which launches a harsh anti-democratic agenda on the basis of his biological theories. The idea subsequently becomes particularly rampant in Nazism, with its racist and eugenic visions of a purified

point, and we find responses from a philosophical point of view in the works of, for instance, Giorgio Agamben, Roberto Esposito, and Antonio Negri and Michael Hardt; in the field of the social sciences the idea of biopolitics has notably been explored by Paul Rabinow and Nikolas Rose, and there is also a growing literature in legal studies, international relations theory, demography, history of medicine, and biology, as well as in many other emerging disciplines. Surveying all the multifarious uses of the term in contemporary research would probably show that they are held together only by a set of family resemblances, although the reference to Foucault is a thread that runs through most of them.

Volkskörper that in return requires a strict *Biopolitik*, as for instance in Hans Reiter's (at the time president of the Reich Health Department) programmatic 1934 address "Unsere Biopolitik und das Auslandsdeustchtum." After a long period of incubation after the war, the theme was reintroduced in the 1960s by Edmund Wilson and others, now in the form of a "sociobiology" that rejects the quasi-mythical ideas of an organicist society and instead draws on evolutionary theory as a general explanation for politics. In all of these theories, if we disregard their obvious differences on the level of ideology and scientific standing, we find both a descriptive and a normative element: politics and social structures are *in fact* based on biological processes, and we *ought* to make this into both an ethical and a theoretical point of departure for our actions, instead of the "culturalism," individualism, and idea of free choice prevailing in the social sciences. Foucault's proposal, when he first introduces the theme of biopolitics in the first volume of the *History of Sexuality,* and in the final section in the lecture series *"Society Must Be Defended"* (both 1976), was rather to subject the first part of this tradition to a genea-logical analysis, although he soon, as we will see, dropped the connection to racism and Nazism and instead came to focus particularly on liberalism, where individual freedom and choice are obviously a central feature, and in this sense the object of his research fits none of the earlier categories. As Thomas Lemke points out, while Foucault's reinvent-tion of the term occurs in the context of the new ecological concerns of the mid '70s, the new discoveries in biotechnology, in vitro fertilization and prenatal testing—all of which spawned a large debate around the impact of the life sciences on politics and society in general, and on whether science needs to be regulated—his historical genealogy shifts the perspective both from a merely technological point of view, and from theories that begin from an already given distinction between life processes and the political; see Lemke, "From State Biology to the Government of Life." For general discussions of the life sciences and the politics that provide a context for Foucault's work, see also Theo Steiner (ed.), *Genpool: Biopolitik und Körper-Utopien* (Vienna: Passagen, 2002) and Volker Gerhardt, *Die angeborende Würde des Menschen: Aufsätze zur Biopolitik* (Berlin: Parerga, 2004). The question for several of the contributors below (above all Oksala, Mills, and Reid) however points to a subsequent step in this debate, i.e. whether we, in the wake of Foucault's historical analyses, can re-invent an *affirmative* idea of biopolitics that would not only amount to a resistance to the earlier right-wing version, as well as the current neoliberal one, but also bear on how to articulate philosophy, science, and politics in a new determination of life. A similar question underlines the work of Roberto Esposito, and perhaps also that of Giorgio Agamben, as well as, in a more general fashion, many of the current investigations being carried out on the vitalist dimension in Deleuze.

While this diverging and even confusing reception undoubtedly is the result of the urgency and complexity of the issue itself—ranging from the impact of the life sciences on philosophy and social theory, to the way in which a "life" situated at the outer limits of our inherited discourses of rights and citizenship imposes itself as a matter for thought—to some extent it also stems from the plurality of perspectives opened up in Foucault's work from the latter half of the 1970s onwards. It is to this period that the present collection of texts is dedicated. This kaleidoscopic break-up emerges out of the crisis that seems to have set in sometime after both *Discipline and Punish* (1975) and the first volume of the *History of Sexuality* (1976), in which Foucault begins to re-orient his research.

This hypothesis of a series of breaks in Foucault's work is what motivates the three-stage picture that we find in, for instance, Deleuze's architectonic reading in *Foucault* (1986), as well as in much of the secondary literature from the '80s and early '90s. In the first phase, roughly from *The History of Madness* (1961) to *The Archeology of Knowledge* (1969), Foucault would have devoted himself to "archeology," where he pursues a Kantian quest for the conditions of possibility of knowledge, although transformed into a search for "historical a prioris"; in the second phase, beginning with *L'ordre du discours* (1970), he would have shifted to questions of "genealogy," where, in the wake of Nietzsche, he connects the archives of knowledge to relations of power, and where discipline, as a technology that fundamentally acts on the body, becomes a key factor for the understanding of modern societies. Finally, there is a period of silence and gestation after 1976, where he is said to have returned to the self as something to be made and con-structed through empirico-historical practices ("technologies of the self"). This returning to the idea of the self, however, was also meant to displace or circumvent the transcendental claims about subjecthood inherited from the tradition of Descartes, Kant and Husserl. This third phase, then, would have finally come together in the last works, volumes two and three of *The History of Sexuality*, *The Use of Pleasures* and *The Care of the Self*, which develop this new idea of a situated and constructed subjectivity mainly on the basis of Greek and Roman texts.

Although obviously not simply incorrect, this three-stage model has the obvious limitation of being based only on the published works. The subsequent edition of the lecture courses from the Collège de France, as well as the massive four-volume *Dits et écrits*—comprising most of (though not all) of his essays, reviews, and interviews—have complicated the image we have of Foucault's development, and can be said to have initiated a substantially new reception. These three phases not only have an inner complexity, but also contain many themes and questions that extend across those divisions: the first must include the early work leading up to the study of madness, and its contradictory and shifting exchanges with phenomenology and existential psychoanalysis extend throughout the 1960s; the second must be seen as a gradual process of discovery of a new idea of power, which however always was in a state of flux; and the third is rather something like a prismatic diffraction, to the effect that we here can see a multiplicity of questions that never coalesce into a unified complex.

In hindsight, the published works appear like snapshots, momentary cutouts from a process; perhaps they were too well written, too obsessed with creating self-enclosed unities, all of which has generated criticisms whose depth of ambition may vary, while the underlying theme remains the same: Foucault's mode of presentation is itself panoptical and totalizing, it mirrors the forms of discipline that he wants to uncover, and it immobilizes us in the face of an irreversible dystopia. Reading the lectures from the Collège de France however provides an efficient antidote to this. In those lectures we see Foucault at work, constantly returning to older questions, restating and reframing them, always prepared to discard his earlier claims if a new angle should present itself.

In this way, the unity of Foucault's work does not consist in a system or a set of theses. Such reductive ideas have become widespread, especially among his critics, where they have petrified into an unshakable *doxa*: the individual, desire, and subjectivity are nothing but surface effects of discipline; the human and social sciences are nothing but the exertion of power; and even truth itself is only a deceptive mask behind which we find rhetoric, or even coercion and simple violence, all of which makes Foucault into little more than a modern Thrasymachus, and an easy prey for all the self-appointed Socrateses of this world. If there is a unity, it must rather be sought on the level of questioning, in the necessity of never remaining satisfied with the answer just given, and of constantly returning to the starting point in order to frame the investigation differently. This incessant mobility might no doubt provoke a certain impatience, not least since it makes Foucault less useful as a quasi-magical reference in debates, either as an authority or a projected opponent, but it is also a salient example of what Wittgenstein once claimed, to be sure in a different context: philosophy is not a doctrine, but an activity.

Discipline and after

The major shift that occurs, first in the 1977–78 lectures on *Security, Territory, Population,* and then in the lectures from 1978–79 on *The Birth of Biopolitics,*[3] is the move away from the idea of discipline as the prevalent structure of modern societies (another important and related change would be the abandonment of the Nietzschean "war" model for social relations in *"Society Must Be Defended"* from 1976–77; see Julian Reid's discussion of this below). It would be too hasty to see this move as already connected to the later work on subjectivation; in fact, after the abandonment of the disciplinary model, it seems as if Foucault increasingly came to distrust any overarching theory. What we find is a spectrum of questions, or multiple guidelines for further research. They intersect and resonate with each other, but also diverge in different directions. Without claiming to be exhaustive, it is maybe helpful to mention at least four of the major issues that traverse his work from 1976 to the end, of which biopolitics is only one, and by no means the predominant one.

1) *Biopolitics.* The concept is first presented in the final chapter of the introductory volume to *The History of Sexuality* (1976), and it also appears in the final section of *"Society Must Be Defended"* from the same year. A "society's 'threshold of modernity' has been reached," Foucault famously says, "when the life of the species is wagered on its own political strategies. For millennia, man remained what he was for Aristotle: a living being with the additional capacity for political existence; modern man is an animal whose politics places his existence as a living being in question."[4] In this conception, biopolitics—or biopower, as Foucault more often says here—is understood as the other side of an "anatomico-politics of the human body," in a way that remains closely connected to discipline. Here, the structure of biopower seems in fact to be a result of, or even an aside within, the genealogy of the sex. Piecing together the various parts of this initial presentation, we can see that biopower has a three-tier structure. On the lower or micro-level it works by individualization, or more precisely by *producing individuality* as the focal point of all the different techniques for monitoring

---

[3] *Security, Territory, Population: Lectures at the Collège de France, 1977–1978,* ed. Michel Senellart, trans. Graham Burchell (Basingstoke: Palgrave Macmillan, 2007). *The Birth of Biopolitics: Lectures at the Collége de France, 1978–1979,* ed. Michel Senellart, trans. Graham Burchell (Basingstoke: Palgrave Macmillan, 2008). Henceforth cited as STP and BB.
[4] Foucault, *The History of Sexuality, Vol. 1: An Introduction,* trans. Robert Hurley (New York: Vintage, 1978), 143.

the body politic, which now fractures into a *living multiplicity* of individuals. In this sense, individuality is produced by those very disciplinary techniques that at the same time *discover* it as their proper object. But this process also makes another object visible on the higher or macro-level, namely *population*, which is how individuals appear when they are treated as statistical phenomena, in terms of collective health and collective forms of reproduction and life. And finally, there is a crucial link between the production of sex as individuating force and the production of sex in relation to the population, or to the collective entity: the *family*. The family is the site of exchange between individuality and collectivity, the relay through which all individuals have to pass in order to become members of the reproductive body politic.

In *Security, Territory, Population* the analysis will however soon take a different direction, and "biopolitics" (which is now the term generally used) merges with the problem of "governmentality," to the extent that Foucault, especially in the subsequent *The Birth of Biopolitics*, almost seems to lose interest in the topic.[5] As we will see, the first and second model of biopolitics do share some features, above all the idea of "population," but there are also important differences.

2) *Sexuality*. Foucault begins to explore this in a way that remains almost exclusively focused on the production of the subject, its inner depth, and its desires through the mobilization of external apparatuses, or *dispositifs*.[6]

---

[5] This might exemplify a move that Foucault alludes to in his very last interview, i.e., that he would often preserve the title of a book or a lecture series although the initial ideas had been discarded. See Foucault, *Politics, Philosophy, Culture: Interviews and Other Writings, 1977–1984*, ed. Lawrence D. Kritzman (London: Routledge, 1990), 64. Foucault himself suggests that the 1977–78 lectures could more fittingly have been called "a history of governmentality" (STP, 108). James Miller cites an anonymous listener reacting to the shift in the 1978–79 lectures: "He could not go on. And it was clear that this problematic, of bio-politics, was over for him—it was finished." See Miller, *The Passions of Michel Foucault* (London: HarperCollins, 1993), 299. Others have claimed that it is only in the analysis of modern neoliberalism that the analysis of biopolitics is completed; see, for instance, Johanna Oksala's contribution below.

[6] The word "assemblage," although already occupied as a translation of Deleuze and Guattari's *agencement*, seems more apt, since it avoids the conflation with "state apparatus." But even though *agencement* and *dispositif* share many features, they are not equivalent. The differences between Deleuze and Guattari's philosophical constructivism and Foucault's analytic of power to some extent reflect the difference in temperament and style between the philosopher and the historian. Foucault's questions bear upon how we have become the types of subjects that we are, but remains largely silent when it comes to positive programs for new types of subject-formation, whereas Deleuze and Guattari seem fascinated with synthetic and universal-historical models, and their project is to discern conditions for the lines of flight that open up in every assemblage.

That an over-emphasis was placed on the idea that sex, as well as all other "deep" features of the self, would be simply produced by the application of external forces (which in turn seems to presuppose the limit-idea of some inert primary matter), is probably one of the reasons for Foucault's crisis and the decision to abandon the project of writing a history of sexuality, doubtless contributing to his long silence; he published no books between 1976 and 1984. It is here that the lectures form such a formidable resource for tracing this long process of reorientation, where biopolitics and the problem of sexuality seem to take off in two different directions. Between 1977 and 1979, Foucault pursues the problem of biopolitics, although in a way that makes no reference to the earlier framework of a history of sexuality; and when, in the lectures from 1980–81,[7] he returns to the techniques for interpreting the subject—which he earlier saw as part of the deployment of sexuality in the modern period—he treats it very differently: the "hermeneutics of the self," whose Greek origins he now traces in great detail, can no longer be understood as a modern avatar of Christian confessional technologies, rather we must see the latter as made possible by a much longer historical development that takes us back to Plato. Sexuality and sex are now integrated in a problem of self-relation that has a much wider scope: a terrain is opened up where once more Foucault takes up a dialog with both phenomenology and psychoanalysis that was begun in the work before *Madness and Civilization*, and where the problem of the constitution of "experience" comes to the fore again.[8]

For Deleuze and Guattari, a society is held together not so much by its segmented forms as by that which escapes such orderings. This points to a proximity to Foucault's idea that resistance comes first. But on the other hand, Foucault remains critical of all non-historical and ontological conceptions of desire as a productive force, which is why he abandons the idea of the "desiring subject"—initially, in the first volume of the *History of Sexuality*, since it appears to him as a product of modern confessional technologies, and then, when he in the second and third volume and the adjacent lectures takes us back to Greece, since the concept requires an even more extensive historical genealogy. For Deleuze's comments on these disputes, see "Désir et plaisir," in *Deux régimes de fous* (Paris: Minuit, 2003).

[7] *The Hermeneutics of the Subject: Lectures at the Collège de France, 1981–1982*, ed. Frédéric Gros, trans. Graham Burchell (New York: Palgrave-Macmillan, 2005).
[8] In Foucault's first major publication, the introduction to his and Jacqueline Verdeaux's 1954 translation of Ludwig Binswanger's *Traum und Existenz*, he is looking for a concept of experience that would incorporate the lessons of phenomenology and psychoanalysis, but also transcend them. Foucault shows how Binswanger takes his cues from both Husserl and Freud, but only in order to transgress both of them towards a different conception of consciousness. In this reading, we can see the beginning of a long-standing and by no means simple critique of the psychiatric establishment, bound up with the idea that madness harbors a profound experience of *limits*, that it has a

13

3) *Enlightenment* and *modernity*. This is where Foucault proposes that we think of his work as dealing with the "history of the present" and the "ontology of actuality." Here we also find his exchanges with Jürgen Habermas, Hubert Dreyfus, and Charles Taylor, as well as his renewed reading of Kant. Accordingly, Foucault picks up the thread from his extensive, and for a long time little known, early work on critical philosophy, comprising the translation of Kant's *Anthropology* in 1961 as well as a long preface (which for a long time remained unpublished), both of which were submitted in 1961 as a *thèse complémentaire* to *History of Madness*, and where we find an outline of what would later become the analytic of finitude in *The Order of Things*.[9] In the later reflections, Kant is however no longer an obstacle for thought that needs to be overcome, but a positive resource. While these discussions were to some extent forced on Foucault by the debate on postmodernity, which was singularly irrelevant to his work, they allowed him nonetheless to situate his own research in the wake of Weber and the Frankfurt School, which he had only rarely commented upon in his earlier writings.[10] What is central, he now says in retrospect, was

---

substance, and not just consists in dysfunction, disorder, and deviation. When Foucault in Binswanger locates a dialectic between experience and institution, or between anthropology and social history, his question is whether we can unearth something like a shared historicity that would be a common root of these two modes of analysis, and bring together the subjective and objective in a third dimension that does not treat them as fixed forms, but can account for their mutual and conflicted emergence. This question will resonate throughout Foucault's work, as we can see in one of the last texts, a sketch for a preface to *The Use of Pleasures*, where he returns to the question of experience, and the common root as one of the underlying themes of his work. See "Introduction," in *Dits et écrits*, Vol. I (Paris: Gallimard, 1994), 65-119, and "Préface a l'"Histoire de la sexualité,'" in *Dits et écrits*, Vol. IV, 578-84.

[9] See Immanuel Kant and Michel Foucault, *Anthropologie du point de vue pragmatique, précédé de Introduction à l'Anthropologie* (Paris: Vrin, 2008), ed. Daniel Defert, François Ewald, and Frédéric Gros. In fact, as the editors point out, the theme is present already in Foucault's first preserved philosophical text from 1952–53, a series of lectures at the University of Lille, under the rubric "Knowledge of Man and Transcendental Reflection." These still unpublished 97 manuscript pages discuss Kant, Hegel, Feuerbach, Marx, Nietzsche, and Dilthey, and form an early matrix for the introduction to Kant's *Anthropology*. I discuss the link between the early and late readings of Kant in more detail in my "Governance and Rebellion: Foucault as a Reader of Kant and the Greeks," *Site* 22-23 (2008).

[10] The Frankfurt School is mentioned briefly in *The Birth of Biopolitics* as one of the outcomes of the dual heritage of Weber in Germany, one line passing through ordoliberalism and the other through Critical Theory, both of which then come together in a violent clash during the events of May 1968: "history had it that in 1968 the last disciples of the Frankfurt School clashed with the police of a government inspired by the Freiburg School, thus finding themselves on opposite sides of the barricades, for such was the double, parallel, crossed, and antagonistic fate of Weberianism in Germany." (BB, 106)

not the "structural analysis of truth," in a line that leads from Kant's first Critique to the first phases of analytical philosophy as well as to the epistemological claims of early phenomenology, but the question of truth as situated, historical, and finite, as a series of shifting horizons that must include the present of the questioner. It is here that Foucault once more encounters the later work of Heidegger (and to a lesser extent Hegel), and the question of what it means for truth to have a *history*, without being simply reducible to empirical conditions, i.e. truth as a series of problems imposed on thought.[11]

4) Finally, there is the theme of *parrhesia* and truth-telling (which is dealt with in Maurizio Lazzarato's contribution below). Foucault once more returns to the Greeks, this time not in order to investigate various forms of self-relation and self-constitution, but instead to consider the role of the intellectual in public life and Greek democracy (significantly, the 1981–82 lectures begin with an extended meditation on Kant's analysis of *Öffentlichkeit* before going back to the Greek material).[12] But although the work on *parrhesia* forms the core of the last lectures delivered before his death in 1984, they cannot be taken as a last will and testament; they remain just as open and inconclusive as the preceding ones. The end itself seems to arrive *in medias res*; the final words from the lecture, dated March 28, 1984, read:

> The truth of life before the true life, it is in this reversal that Christian asceticism fundamentally modified an ancient asceticism that always aspired to lead, at the same time, the true life and life of truth, and that,

[11] The ultimate question that Foucault proposes in his analysis of the history of sexuality (most explicitly in the introduction to the second volume, *The Use of Pleasure*) bears on how man has constituted himself as at once a subject and an object, explicated his relation to himself, and opened up a space of self-reflection in a "truth game." Such a "hermeneutics of the self" does not concern knowledge in the sense of propositional truth, but the constitution of a particular type of experience of the self as a *problem*. This history of truth intersects Heidegger's conception of the history of metaphysics as a series of finite horizons within which beings can be given in experience, and for Heidegger too, truth does not refer to mental representations or propositions as correlated to states of affairs, but to a pre-objective and pre-subjective "openness" for all types of subject and object positions, *aletheia* as the clearing in which beings can be encountered. But unlike Heidegger's meditations on metaphysics, which claim to locate those moments in the history of metaphysics where beings as such are given in new ways, the perspective opened up by Foucault's rereading of the *dispositifs* of sexuality and the analyses of processes of "subjectification" and "technologies of the self" wants to allow an open multiplicity of mundane practices to play a constitutive role, although without giving in to a simple historicism.

[12] *Le gouvernement de soi et des autres: Cours au collège de France, 1982–1983*, ed. Frédéric Gros (Paris: Gallimard/Seuil, 2008).

at least in Cynicism, affirmed the possibility of leading such a life of truth. But listen, I had more things to say about the general framework of these analyses. But it is too late now. So, thank you.[13]

Whether these developmental lines (to which others undoubtedly could be added) may be knitted together at some more profound level must remain an open question. It is indeed true that certain themes, like the problem of governing, recur regularly, although their significance seems to vary contextually; other topics, like sexuality, suddenly return after long absences, as if all of these questions would co-exist at a deeper level, although not simply in a peaceful and thus mutually indifferent way, but as a series of unresolved tensions. If one were to seek out a pervasive theme, then the idea of "governing" would probably be the most likely candidate: on the one hand, we are presented with the governing of others, the roots of which take us back to the Christian pastorate and, specifically, to the relation between the shepherd and his flock; on the other hand, we also encounter the governing of oneself, which becomes a question of the subject and its self-relation as *ethos*. The success, even inflation, of the concept of "governmentality" in subsequent political science (addressed in Thomas Lemke's contribution below) would testify to the fecundity of choosing this angle, although it would be far too reductive if we were to allow it to subsume all of Foucault's research since the mid 1970s, at the expense of other concepts like truth, subjectivity, and experience. These remain just as pertinent, indicating that the question of interiority and individual experience, with which he began, by no means disappears, but is rather resituated within a more encompassing analysis of historical practices.

What position, then, does biopolitics hold in this complex development? The extended historical analyses of the history of governmentality initiated in *Security, Territory, Population* seem to push biopolitics aside, and the topic only recurs briefly at the end of the volume; and despite its title, the next lecture series, *The Birth of Biopolitics*, after having summarized the results of the previous analyses, moves ahead to discuss modern liberalism, whose connection to the earlier theme is at least not obvious (an argument for the continuity of these two discussions is given in Johanna Oksala's contribution below). A reasonable conjecture would however be that these moves constitute different steps in a gradual slide from discipline to subjectivation, a process in which the problem of biopolitics may be taken

---

[13] *Le courage de la vérité: Le gouvernement de soi et des autres II, Cours au Collège de France 1984*, ed. Frédéric Gros (Paris: Gallimard/Seuil, 2009), 308f.

to form an intermediary stage that was essential in redirecting the investigation, and yet remained too closely affiliated with a conception of power that tended to preclude, or at least downplay, processes of self-formation—whence the emergence of an "ethical" turn in his last work.[14]

Such a reading, which proposes to locate a gradual shift towards the problem of individual existence, is in one sense supported by the two final published volumes on sexuality, *The Use of Pleasure* and *The Care of the Self*, where—to the extent that it remains at all—the question of the *bios* bears on inwardness and selfhood. But on the other hand, the lectures from the same period, *Le gouvernement de soi et des autres* and *Le courage de la vérité*, take the opposite route, initiating enquiries into the public, into the political role of truth, of democracy, and, moreover, an account of the figure of the intellectual as a truth-teller. In this sense, the question of the place occupied by biopolitics in the trajectory of Foucault's last work may be futile, since what we get is not a movement that gradually integrates questions that at first seem unrelated; we encounter instead a multiplicity of outward paths.

Security and freedom

What is strikingly new in the 1977–79 analysis of biopolitics, when compared to the preceding claims in *The History of Sexuality*, is the privileged status accorded to the liberal tradition. If liberalism earlier had been treated largely as a surface phenomenon—in the sense that the autonomous individual as a bearer of rights, capable of rational choices, was implicitly understood as a product of processes of discipline—this freedom now appears as a correlate to what Foucault, in *Security, Territory, Population*, calls "apparatuses of security." It is understood as part of a new type of governing for which life is a reservoir that must be tapped into rather than subjected to legal or disciplinary strictures.

At the outset of *Security, Territory, Population*, Foucault explains this by analyzing the example of theft. Theft can be understood as an infraction that must be dealt with according to a predetermined scale of punishment,

---

[14] We now posssess another important text that would allow us to reconstruct Foucault's trajectory, *Du gouvernement des vivants: Cours au Collège de France (1979–1980)*, ed. Michel Senellart (Paris: Seuil, 2012), which however appeared too late to be taken into account here. The following lectures from 1980–81, *Subjectivité et Vérité*, still remain to be published.

i.e. as a *juridical* problem; it can be seen as a form of deviant behavior that must be corrected through the application of various techniques, i.e. as a *disciplinary* problem; and, finally, it can be taken as a statistical phenomenon, where one must balance the losses and gains of disciplinary measures, and perhaps even allow for a certain latitude of crime, according to which the problem is formulated in terms of *security*. This third solution is based on probabilities, on a calculus of cost. The problem it poses is how to attain an optimal balance. Foucault cautions us against seeing these three models as a chronological development from the archaic to the modern, or as constituting a path towards increasing rationality, instead they are always co-present as complex structures where, in each case, one of the elements exercises dominance over the others. Thus, for example, security integrates the juridical and the disciplinary, but in a subordinate form, just as the juridical and the disciplinary contain the other moments.

In the following lecture Foucault proceeds by studying other aspects of the new "apparatus" (*dispositif*) of security: its spatial dimension, the implications of chance and event, techniques of normalization, and finally the connection between security and population, which he claims to be the fundamental discovery of eighteenth century political thought. Here it may suffice to point to a few basic traits. Sovereignty, Foucault says, is exerted over a *territory* and a multiplicity of political subjects; discipline is applied to singular *bodies*, to their affects and passions, and the grid "individual" is both a technique of application and an intended result; security, finally, relates to the *population* and its inherent dynamic, as a living entity following laws that politics must obey, and in this sense the population constitutes the *physis* of politics. In all three cases we find a pervasive spatial implication, a point borne out by Foucault's use of the *city* as a common denominator (for more on Foucault's relation to architecture and urbanism, see both Łukasz Stanek's and Helena Mattsson's contributions below). The first example is a text by Alexandre Le Maître, *La métropolitée* (1682),[15] which provides the spatial schema for sovereignty. Le Maître conceives a three-part structure, corresponding to the division of peasants, craftsmen,

---

[15] Very little is known about Le Maître—who in fact, contrary to what Foucault says, seems to have been one of the first to use the term "population"—except his two works, the above cited *La Metropolitée, ou, De l'établissement des villes capitales, de leur utilité, passive & active, de l'union de leurs parties & de leur anatomie, de leur commerce, etc.*, and *Oeuvre de Troye ou de l'excellence et de l'ancienneté des fortifications demontrées par les modèles de la nature, les passions* (1683). For discussions of the context of these texts, see Claude Lévy, "Un plan d'aménagement du territoire au XVIIe siècle: 'La Métropolitée' d'Alexandre Le Maître," *Population*, Vol. 12 (1957): 103-114.

and city administrators, and then in turn to countryside, city, and capitol. This is a circular model of power, where knowledge and the radiance of superiority emanate from the center outward, so that the economy, territory, and sovereign power are coordinated and superimposed. The second example is the city of Richelieu (Foucault also mentions Oslo and, somewhat enigmatically, Gothenburg, as other possible cases), which becomes the model for the disciplinary city. Built from scratch with the Roman military camp as its model, it provides us with a basic geometric figure that is then divided up into smaller subfigures. Unlike Le Maître's elaborate correspondences between micro- and macro-cosmos, the geometrical analysis applied to Richelieu does not express a transcendent order, but becomes a tool for the production of a disciplinary space, for the ordering of manifolds. Discipline, Foucault says, is a radical construction of an idealized reality that runs parallel to everyday life. We can here see a significant shift from *Discipline and Punish*, where discipline was applied to physical bodies and operated through a segmentation and analysis of actual space: in the later lectures discipline becomes something *ideal*, which makes it possible to oppose it to a security that operates in the *real*.

The third example, which introduces security as a problem of urbanism, is Nantes, a city based on trade, with the possibility of economic growth as its guiding idea. The city becomes an instrument for controlling circulation in all its aspects; it is part of a larger network comprising the surrounding countryside as well as other cities, so that the possibility for future development is inscribed into the plan from the very start. If discipline operated in an empty abstract space to be constructed from scratch, as in Richelieu, security could be said to work with a set of fluid conditions, constantly fluctuating quantities, and future probabilities. The task of security is to invent a multifunctional order, and to calculate the negative and positive outcome of any given measure; it does not apply to a fixed state, but relates to a series of future events. If sovereignty monopolizes a territory and locates a central command, and discipline structures a space and sets up a hierarchy, then security attempts to plan an environment or "milieu" in relation to a set of possible events.

The city, Foucault says, becomes the paradigmatic place for the articulation of the problem of the "natural" quality of humankind in an artificial milieu. Such a problematic comes across, for instance, in Jean-Bapiste Moheaux' *Recherches sur la population* (1778), and which Foucault locates as one of the first explicit texts on biopolitics. Sovereignty does not disappear, just as little as discipline would simply be displaced by security. It undergoes instead a decisive mutation, so that it now is exercised over a nature inaccessible to legal and disciplinary measures, but nonetheless exercises influence over man's body, soul, and moral character.

Discipline is centripetal, it isolates spaces and creates segments, it focuses and encloses, all of which can be seen laid out in great detail in Nicolas Delamare's *Traité de police* (1705–19); apparatuses of security, on the other hand, are centrifugal, and they aim to integrate new things in ever widening circuits. Discipline strives toward a regulation of details, whereas security at a certain level allows things to run their course. Discipline, Foucault says, divides things into licit and illicit, and to this extent it is based on a law that is to be increasingly specified. In law, order is what is supposed to remain once everything prohibited and disorderly has been removed, and this is intensified in discipline, since it also tells you *what* to do, which is why the convent can be taken as its ideal form. In all of this, we can undoubtedly detect an important shift away from the earlier work, where the "juridical" conception of power, based on binary divisions, was *opposed* to the attention to detail and modulations in discipline; here they sometimes appear as two stages of the same process, the counterpoint to which would now be security.

Deploying the apparatuses of security no longer means to exercise sovereignty over subjects, but becomes what the Physiocrats called a *physical* process, within which exhaustive control no longer is an issue. In this way, Foucault suggests—and here he once more significantly modifies his earlier theses—the Panopticon, in which the central tower and its possibility of continual inspection is what displaces the discontinuous violence of the sovereign, appears as an archaic rather than a modern model. In the apparatuses of security, the *modus operandi* is not panoptic *surveillance*; it is rather to take a step back and *observe* the nature of events—not in order to attain some immutable essence of things, but to ask whether they are advantageous or not, and how, moreover, one can find support in reality itself, making it possible to channel them in an appropriate direction. In this respect we can say that the law operates in the *imaginary*, it imagines something negative; discipline is applied in a sphere which is *complementary* to reality; security, finally, "tries to work with reality, by getting the components of reality to work in relation to each other" (BB, 47) —which is what the Physiocrats meant, Foucault suggests, when they said that economy in fact is a physics, and that politics still belongs to nature.

This, Foucault concludes, is the basic tenet of what would become liberalism, and it allows for a different understanding of the role of the idea of

FOUCAULT, BIOPOLITICS, AND GOVERNMENTALITY

freedom in the development of capitalism. Correcting his earlier proposals, which seemed to suggest that the emergence of the idea of freedom and of liberal ideologies must be understood in connection with the massive deployment of disciplinary techniques (the free individual being a result of processes of discipline), Foucault now claims that freedom should be located in correlation to the apparatuses of security, in the sense that such apparatuses could only become operative if one assumes a certain freedom, a freedom of circulation, lodged in that "political nature" within which they are to be deployed.

Liberalism, Foucault proposes, is thus first and foremost not an ideology—in the sense of a false, distorted, or imaginary representation of reality—but a technology of power, or a way to work *with* reality; liberalism, seen from the perspective here adopted—which does not preclude other perspectives, although it implicitly claims to *precede* them—does not provide us with a theoretical and/or ideological smoke-screen behind which other and more real things (actions, practices, material events) are taking place; instead, itself a practice, it is a way to make certain things real by working with, intensifying, tempering, or redirecting processes already underway in reality itself.

Versions of liberalism

In the following lectures, *The Birth of Biopolitics*, Foucault begins by summarizing his earlier investigations of eighteenth-century discourses, but then surprisingly makes a long jump, devoting the major part of the following to discussions of liberalism from the twentieth century, first focusing on German ordoliberalism, and then, after a shorter digression on France, on the neoliberal theory of the postwar Chicago School.

It has often been noted that this is the only text (the interviews apart) where Foucault comments at any length on the present. The shift of perspective also has effects at a stylistic level, and Foucault has never come so close to writing history of ideas in a traditional sense; often he seems to be simply presenting us with a series of intellectual biographies, drawing on standard works, as if he were trying to complete an inventory of a field rather than providing his own interpretation.[16] It appears as if Foucault is

---

[16] See Mike Gane's review of *Security, Territory, Population* and *The Birth of Biopolitics*, in *Theory Culture Society*, Vol. 25, No. 7-8 (2008): 356f, and Foucault's own comments, BB, 102f ("I will break away from my habits and give a few biographical details…").

unsure about how to handle this new and pristine matter, and he approaches it with a mixture of curiosity and bewilderment. When he discusses the eighteenth century, there is always a framework derived from his earlier archeological and genealogical analyses, which can be enriched, reconstructed, and set in a new light, and the overlay of successive interpretations provides us with a sense of depth. A salient case of this would be the analysis of the concept of Man: in *The Order of Things*, Man emerges in and through the epistemological break that separates the classical from the modern age, as a new dense figure of finitude relating to itself; in *Discipline and Punish*, Man is as it were precipitated out of the techniques of discipline and interrogation that provide the deviant with a "soul," which in turn becomes the ground and source of verification for these techniques; finally, in *Security, Territory, Population*, the discourse on Man "should be understood on the basis of the emergence of population as the correlate of power and the object of knowledge" (STP, 79). The discussions of twentieth-century texts however lack such frames, which is perhaps also why we here come close to a kind of self-problematization and an attempt on Foucault's part to determine the position from which he is speaking in a very concrete sense. "I have not made these analyses just for the pleasure of engaging in a bit of contemporary history," he claims (BB, 179), but he provides us with no clear statement as to what this ulterior purpose might be.

If we would explain this shift by situating the lectures as responses to their immediate context, the crisis of Marxism and the sudden visibility of neoliberal theories in France would be the most likely reference points. Textual references would include, for instance, Henri Lepage's highly publicized *Demain le capitalisme*, from 1978, and which in fact seems to have been one of Foucault's major sources, as well as the emergence of the "new philosophers" (it should be noted, parenthetically, that Foucault reviews André Glucksmann's *Les maîtres penseurs* favorably when it appears in 1977).[17] Others have pointed to Foucault's misreading of the Iranian revolution as a reason for his reassessment of certain parts of Enlightenment political theory,[18] and even to events on a purely personal level.[19] But what-

---

[17] See Foucault, "La grande colère des faits," *Le Nouvel Observateur*, May 9, 1977: 84-86.
[18] Foucault's articles in *Corriere della Sera* and *Le Nouvel Observateur* in the spring of 1978, in the immediate aftermath of the Iranian revolution, have been the object of long discussions. For some they are the result of a regrettable misreading of the stakes, which Foucault however soon corrected; for others they are the symptom of a political romanticism in the search of something absolutely other behind modern parliamentary democracy, which has sometimes earned him the comparison to Heidegger in 1933. In an interesting analysis, Alain Beaulieu has suggested that we must see this engagement as an

ever the contextual or personal reasons may be for this shift, the problem obviously remains as to what these lectures signify from within the trajectory of Foucault's thought, i.e. to what extent can they be taken as a response to a proper Foucauldian problem.

When Foucault suggests that liberalism is an exemplary form of governmentality for the exercise of biopolitics, his claim is that it is opposed to the police, inherent in the doctrines of Raison d'Etat, and that it discovers a new art of limiting the interventions of the state.[20] In liberal government the power of the state ceases to be a goal in itself, which means that it will counter the "Reason of the State" with a "Reason of the smaller State." This art of self-limiting points to a new form of political rationality, the basic question deriving from which will be how to achieve maximum efficiency through minimum intervention. It is also the founding intuition of political

integral part of Foucault's path: at the time Foucault was searching for what he called a "political spirituality" (spritualité politique) supposedly lost in the Western tradition since the Renaissance, which he thought could be found in Khomeini (who "is not a politician," Foucault claims). The catastrophic result of the ensuing events forced Foucault to reconsider these claims, Beaulieu suggests, which may account for the rather different tones in the 1978–79 lectures. See Alain Beaulieu, "Towards a Liberal Utopia: The Connection between Foucault's Reporting on the Iranian Revolution and the Ethical Turn," *Philosophy and Social Criticism* Vol. 36, No. 7 (2010): 801-18. For a broad study of Foucault and Iran, see Janet Afary and Kevin B. Anderson, *Foucault and the Iranian Revolution: Gender and the Seductions of Islamism* (Chicago: University of Chicago Press, 2005).

[19] See Arpád Szakolczai, who refers to an almost fatal accident that Foucault suffered in July 1978; see *Max Weber and Michel Foucault: Parallel Life-Works* (London: Routledge, 1998), 243.

[20] The status of the "newness" of this type of governing is a problem it shares with many of Foucault's analyses from this period, where he becomes more and more suspicious of historical and epistemological breaks and discontinuities, and seems to be constantly pushing the emergence of the phenomenon under scrutiny back in history. Thus, we can detect a symptomatic shift in the sense of "governmentality" as the lectures progress: when the concept is first introduced in *Security, Territory, Population*, it seems to denote an idea of governing that we find sketched out already in the sixteenth century, but whose development was first blocked by the doctrines of Raison d'Etat, so that it only subsequently, from the eighteenth century onward could begin to merge with police and biopolitics, and become a pervasive feature of the modern state. As the lecture series progresses—in fact immediately in the subsequent lectures that take us back to the contrast between the idea of the pastor in early Christianity and model of "weaving" in Plato's *The Statesman*—and even more so in the following lecture series from 1978–79, the term generally seems to denote any form of governing and ceases to have a specific historical reference, and in this sense it could be replaced with "government" or "governing" without any significant semantic loss. The persistent idea of governmentality as a specifically modern phenomenon is probably due to the early separate publication of the lecture from *Security, Territory, Population* where Foucault introduces the topic; see Graham Burchell, Colin Gordon, and Peter Miller (eds.), *The Foucault Effect: Studies in Governmentality* (Chicago: University of Chicago Press, 1991).

economy, which displaces the idea of an external limit on state intervention through law or through an appeal to natural rights, and instead looks for laws that are more akin to natural laws than to moral or legal institutions. This limitation is a "de facto limitation" (BB, 10), and transgressing it does not mean that the government becomes illegitimate and that its subjects would be released from their duties, only that it is a "clumsy, inadequate government that does not do the proper thing" (ibid).

Liberal forms of governing will consequently have a new relation to truth: the market will be the place of verification, which means that political theory finds itself subordinated to a new body of knowledge, namely "economy," a term whose meaning now shifts into a recognizably modern sense—which does not mean that economy here would finally pass over some threshold of scientificity, only that it comes to be endowed with the function of a "truth," itself a part of a new complex of power and knowledge, to which politics is subjected. Foucault's neutralization of the difference between science and ideology in favor of the more neutral *savoir* here remains fully operative: "Politics and economy," he says, "are not things that exist, or errors, or illusions, or ideologies. They are things that do not exist, and yet which are inscribed in reality and follow under a regime of truth dividing the true and false" (BB, 20).

As a consequence of this shift, the transactions and exchanges in "society" can now begin to be distinguished from the machinations of the "state," so that society eventually becomes opposed to the state, and within certain measure—whose extension must be calculated, and is at the core of the new problem—must be left to itself in order to achieve maximum efficiency. This is the emergence of "civil society" as described in Ferguson's *Essay on the history of civil society* (1767), or later, from a different point of view, in the sections on "bürgerliche Gesellschaft" in Hegel's *Philosophy of Right* (mentioned in passing at the end, BB, 309).

As we noted earlier, when the idea of biopolitics was first introduced at the end of the first volume of *The History of Sexuality* in 1976, the concept of population still seemed to hinge around the idea of a police-like control, a power exercised in a top-down manner through various decrees and administrative measures emanating from the state, which would be fundamentally opposed to what is normally perceived as the basic tenet of liberal theory. When Foucault later links biopolitics closely to liberalism, this is not because he necessarily disputes the traditional understanding of the latter, nor because he simply subscribes to it; with regard to the history of political doctrines, or any of the grand concepts of political theory (the state, nature, rights, etc), his method is "nominalist," as he said on many occasions.[21] The point here is that the doctrine of liberty, when seen within the strategic field of political economy, is a way to extract utility, a material and intellectual surplus value, from the individual, or rather, to extract this value through the individual as a grid for the interpretation and governing of reality. In biopolitical terms, this means that the activities of the state will be related to a "life" that always precedes and overflows it, and where this surplus has its origin. On this level there is no contradiction, rather a strategic complementarity, so that freedom (the spontaneity of acting that must be left to itself) and the deployment of apparatuses of security (which themselves include and even multiply disciplinary technologies) increase and reinforce each other: the individual can be discovered as the locus and source of rights and actions, as a new type of political subject that must be given a calculated latitude in order for there to be an increase in productivity. Such a situation comes about through the involuntary interplay of freedoms—the doctrine of "laissez faire," which as its correlate has an "invisible hand" that guides them.[22]

This is the form of governing that provides the impetus for modern industrial societies, and Foucault underlines that we are still within its grip. After the initial summaries of the preceding lecture courses, Foucault then makes a leap into the twentieth century, as if to demonstrate the continued relevance of his earlier discussion; the transition however remains somewhat abrupt, and whether the move into the present is an aside that leads him astray, or in fact provides the ultimate verification of the earlier historical analyses, is a matter of dispute.[23]

---

[21] See for instance "Questions of Method" (1980) in Burchell, Gordon and Miller (eds.) *The Foucault Effect*, where he speaks of "the effect on historical knowledge of a nominalist critique itself arrived at by way of a historical analysis" (86).

[22] In his analysis of Adam Smith, Foucault rejects the quasi-theological interpretation often given of the invisible hand, and instead stresses its invisibility, which points to the "naturally opaque and non-totalizable quality of economic processes" as a way to "disqualify the political sovereign" (BB, 282-83); economics is a "discipline without God" (282).

[23] We could note that the whole of the nineteenth century and all of neoclassical economics disappear from Foucault's view, and Keynesianism and the welfare state are treated solely as enemies of liberalism, which provides a picture that is far too selective even if liberalism as such is the main issue. For a critique of these elisions, see Francesco Guala's review of *The Birth of Biopolitics*, in *Economy and Philosophy* 22 (2006). On the other hand, Guala notes, "economics looks more like a Foucauldian discipline now than it did when these lectures were delivered at the Collège de France" (439). See also the comments by Tiziana Terranova, "Another Life: The Nature of Political Economy in Foucault's Genealogy of Biopolitics," *Theory Culture Society*, Vol. 26, No. 6 (2009): 247. In fact, the welfare state, especially in its Social-Democratic versions, in many respects

The two main cases that he discusses are German ordoliberalism and the Chicago School.[24] The German tradition derives its name from the journal *Ordo*, which began in 1948 as a forum for debate on the postwar reconstruction, but which actually can be traced back to certain intellectual movements from the '30s, most notably the Freiburg school. The immediate context for the ordoliberals was the perverse Raison d'État of Nazi Germany, but they also leveled harsh criticism at the expansive state of the New Deal and Beveridge Plan in England. After the war, ordoliberalism would become a fundamental source for the postwar *Wirtschaftswunder*, especially in its emphasis on the interplay of market and legal and institutional structures. On this point it can be contrasted with the later Chicago school, which would opt for a much more pervasive and radical market perspective, encompassing (in principle) all strands of life.

Unlike much earlier liberalism, where the state had as its primary role intervention in order to mitigate the consequences of the market, the ordoliberals suggested that the role of the state was to ensure the permanence of competition; economic rationality was said to be the antidote to social dysfunctions: "One must govern for the market," Foucault summarizes, "not because of the market" (BB, 121). The enterprise was the founding model for society, and competition replaced the traditional social bond in a "formal game between inequalities" (120) whose only rule was that no player should be allowed to lose everything and be altogether excluded from the game, thus the assurance of a certain existential minimum. The market is however as such a fragile construct, which is why it needs support from state institutions, above all in settling legal conflicts, but also in many other ways: in fact, rather than a state reduced to an absolute minimum, the ordo-

---

seems like a more apt object of study than Anglo-Saxon liberalism, especially in the way it attempts to balance the need for individual agency and centralized political systems by a whole gamut of highly technical governmental strategies. For a discussion of the emergence of the Swedish system in this perspective, see Helena Mattsson and Sven-Olov Wallenstein (eds.), *Swedish Modernism: Architecture, Consumption and the Welfare State* (London: Black Dog, 2010).

[24] While these two cases undoubtedly can be taken as two major sources of twentieth century liberal theory, the motif for choosing the German example, as the editor Michel Senellart points out in his postface (BB, 328), is also what Foucault perceives as an "inflationary" critique of the state (187)—which itself was largely inspired by his own earlier work—that always sees micro-fascisms at work in each of its operations, and was particularly dominant at the time, as in the debates around German and Italian terrorism, the German *Berufsverbot*, and in France surrounding the quarrel over the extradition of the RAF lawyer Klaus Croissant. "Liberty," Foucault says in a phrase that must surely have bewildered many of his listeners, "in the second half of the twentieth century, well, let's say more accurately, liberalism, is a word that comes to us from Germany" (22).

liberals envisage a "permanent and multiform interventionism," understood as the "historical and social condition of possibility for a market economy" (160). This is why, Foucault proposes, "neo-liberal governmental intervention is no less dense, frequent, active, and continuous than in any other system" (145). It will, for instance, require new legal instruments and institutions, and various mechanisms of control that exist in the interstices between state and market, which is why it is not so much a question of the state's sphere of influence increasing and decreasing as if in a zero-sum game, but rather of the state as "the mobile effect of a regime of multiple governmentalities" (77).[25]

In the Chicago school, the market is no longer understood as a historical construct dependent on the state, instead it appears as a natural process, and the model of enterprise is taken much further, so that it comes to encompass the whole sphere of subjectivity, affectivity, and intimacy. The entrepreneurial relation enters into the self, and via the idea of "human capital" the individual's entire behavior, the body as genetic capital, education as investment, marriage, love, and child rearing, can be understood in terms of investment and revenue (see for instance the discussion of the mother-child relation, BB, 229f and 244f). Accordingly, in this variant, neoliberalism becomes a permanent critique of any state activity from the point of view of economic rationality.

Foucault underlines the inventive character of this entrepreneurial, enterprise-like self, and he opposes it to the relative lack of imagination on the Left—there is "no governmental rationality of socialism," he exclaims at one point (BB, 92), although we are not told whether this is either a structural or simply an empirical deficiency. Many readers have been struck by the neutral, even curious tone in Foucault's lectures (although this seems

---

[25] Interestingly, many contemporary debates on globalization similarly tend to see the nation-state in the perspective of such a zero-sum game, as if it must either gain or lose power in an absolute sense; in fact, it is much more appropriate to say that the state is often that which drives globalization by opposing certain parts of itself to others, and by creating new bodies that propel it in new directions. Rather than a fixed entity, the state is perhaps more fruitfully understood as an "assemblage" in the sense of Deleuze and Guattari, as has been proposed by Saskia Sassen in a series of influential books. For a recent synthesis of her work, see Sassen, *Territory, Authority, Rights: From Medieval to Global Assemblages* (Princeton, N.J.: Princeton University Press, 2006). When Foucault suggests that the "state is not a universal nor in itself an autonomous source of power," but a result of "incessant transactions which modify, or move, or drastically change, or insidiously shift sources of finance, modes of investment, decision-making centers, forms and types of control, relationships between local powers, the central authority" (BB, 77), he comes close to this conception.

like a sign of bad faith, since such a neutrality and distance is almost a trademark of Foucault's treatment of historical material). But occasionally he also seems to be overtaken by fascination, which has led some to claim that there might be a liberal turn in Foucault's late work. In a passage that begins as a gloss on, or perhaps even a paraphrase of Hayek,[26] Foucault slips into what seems like an imperative mode and calls for a transformed political imagination: "It is up to us create liberal Utopias, to think in a liberal mode, rather than presenting liberalism as a technical alternative for government. Liberalism must be a general style of thought, analysis, and imagination." (BB, 219)

While the idea of a pervasive turn towards liberalism seems overstated, especially bearing in mind that Foucault, his shifting commitments in practical politics notwithstanding—or perhaps precisely because of their local and specific quality—throughout his career refused to engage in normative political theory, another problem nonetheless remains: the fate of the idea of resistance in the later work, and a question about the extent to which the analysis of neoliberal governmentality at all allows us to assume a critical distance to the present. It has sometimes been claimed that the turn towards the construction of individuality, an "ethic" or even "aesthetic of existence," as Foucault sometimes called it, even though it is developed mainly on the basis of Greek and Roman texts, in fact fits rather smoothly into the kind of entrepreneurial image of the self of neoliberalism.[27] If power in neoliberal societies is no longer exercised through normalization, but through diversification and individualization—or, as Deleuze suggests,[28] in a way that transcends the individual as an entity still too substance-like and inflexible, towards the "dividual," where discipline as a fixed mold is replaced by a continual modulation and control in an open territory—would it not be possible to say that Foucault simply duplicates the power structures of the present? The extent to which Foucault's lectures on bio-politics enable us to extricate ourselves from this scene must remain an open issue. And beyond the specific question of the motifs underlying his

---

[26] Senellart's editorial footnote (BB, 234) points to similar formulations in the postscript to Hayek's *The Constitution of Liberty*, "Why I am not a Conservative." For Alain Beaulieu, this is Foucault speaking on his own behalf; see Beaulieu, "Towards a liberal Utopia," 812.

[27] See for instance Louis McNay, "Self as Enterprise: Dilemmas of Control and Resistance in Foucault's *The Birth of Biopolitics*," in *Theory Culture Society*, Vol. 26, No. 6 (2009): 55-77.

[28] See Deleuze, "Postscript on Control Societies," in *Negotiations, 1972–1990*, trans. Martin Joughin (New York: Columbia University Press, 1995).

reading of the liberal tradition in the late '70s—carried out in a political conjuncture defined by the decline of Marxism, the rise of a new capitalist self-confidence, and a growing suspicion against the state (which is no longer ours, even though we perhaps can be said to live in the last phase)— there is also the question surrounding what a critical theory of the present at all means, and whether a mere genealogical accounting of our ration- alities is sufficient, or if, conversely, radically transcendent, utopian motifs must be introduced: in short, such questioning coheres around what Foucault called the "ontology of actuality."

The contributions in this volume

The following contributions can be divided into three sections, dealing in turn with: A), the governing of life, and if there is an ontology of life that underlies modern governmentality; B), the problem of spatial articulation and how artistic production can be said to be shaped by biopolitics, both on the larger scale of territorial assemblages and on the level of imagination and affectivity; and, C), finally how concepts of agency and subjectivity can be reworked on the basis of Foucault's work, in relation to Lacanian psychoanalysis and to the theory of political subjectivation developed by Jacques Rancière.

A) Governmentality and the ontology of life

In the first contribution to this volume, Thomas Lemke surveys the terrain of contemporary studies of governmentality—a concept that in Foucault's lectures in fact tends to displace that of biopolitics. Lemke points to three fundamental problems that need to be addressed, and which have their roots in Foucault's own work: the first is the idea of a historical succession of sovereignty, discipline and government; the second takes as its focus a set of limitations surrounding the analysis of programs and the role of failure in the study of governmentalities; and the third relates to how politics, materiality, and space are to be better conceived.

Even though Foucault stressed that different modes of power overlap, Lemke suggests that there is a tendency to see sovereignty, discipline, and governmentality as steps in a gradual rationalization of government, which reduces the persistence of violence and repression in contemporary politics, as well as the role of expressive and emotional factors. Secondly, govern- ment programs are often seen as totalizing entities, and contestation against them as residual and marginal, which underestimates the dynamic of resistance, just as it overlooks that programs always contain fissures and inconsistencies. Failure can in this sense be taken not as a clash with reality, but as the very condition of existence for such programs (as Foucault shows to be the case with the prison system in the nineteenth century). The third problem is that the reluctance to engage in a purely negative critique often leads to a complete lack of an evaluative perspective, and therefore to a "technical" theory that duplicates its object of study; inversely, governmentality studies have been reluctant to integrate analyses of technical and non-human networks. Finally, the focus on the territorially sovereign nation state, and particularly Western liberal societies, tends to underestimate global developments and exclude thereby the possibility for the theory itself to be altered by the inclusion of non-Western cases. All of these problems notwithstanding, Lemke however locates a specific strength of governmentality studies in their very heterogeneity and diversity. He concludes by suggesting that the above problems can be overcome by a closer connection to postcolonial theory, gender studies, and science and technology studies.

Johanna Oksala discusses neoliberal governmentality as a specific political ontology. *The Birth of Biopolitics*, she argues, should be interpreted neither as an historical account of the rise of neoliberalism nor as an instance of ideology critique, but rather as an analysis of how neoliberalism constructs a particular kind of reality, with a particular regime of truth, with its own modes of power and subjectivity. The Left, she suggests, has in a particular way been defeated by "truth," since any kind of extra-systemic critique today appears as wholly irrational.

Neoliberal governmentality must be seen as both a continuation and intensification of earlier biopolitics—the health of the markets implies the health of the population—and, along with a new way of exercising power, it also produces a new type of subject, with an entrepreneurial relation to the self, extending throughout all the spheres of experience. Any effective resistance to this regime, Oksala concludes, must therefore question the traditional instruments of politics and proceed along all three axes of "truth," "power," "subjectivity," if, that is, it is to fundamentally change the structure of our present governmentality.

Catherine Mills notes how the increasingly divergent accounts of biopolitics threaten to dilute its critical power. The concept should be preserved, she argues, but only on the condition of further clarification that bears on what the prefix "bio" signifies. As Foucault stated, life always exceeds the way in which it is governed and administered in political techniques, and an affirmative biopolitics must start by acknowledging this. Mills suggests that there are conceptions of life within contemporary theoretical biology that may be productively utilized. This viewpoint places her in opposition to Giorgio Agamben, for whom all biological conceptions of life as necessarily part of a political order.

Drawing on the work of Roberto Esposito and particularly Georges Canguilhem, Mills conceives of the life process as a production of "vital norms" whose purpose is to uphold functioning relations between an organism and its environment. Human environments may effectively be social, nonetheless vital and social norms—while both inseparable and qualitatively different—continually condition one another. Precisely, it is in this very conjunction—according to which vital norms provide "errors" that concomitantly the biopolitical state reacts to and tries to manage— that the connection between life and politics must be sought.

Julian Reid too is committed to an ontology of life as a condition for an affirmative biopolitics, but approaches it from the point of view of war, returning to Nietzsche as a source for a conception of life as constituted by conflict. War, Reid suggests, cannot be evacuated from an understanding of life, the problem is rather how we can play different conceptions of war against each other in order to find a way to resist neoliberal versions of biopolitics.

Particularly relevant here is Foucault's self-critique in *"Society Must Be Defended,"* where he rejects the model of war, and the subsequent theories of Roberto Esposito, who (similar to Catherine Mills) draws on Foucault's preface to Canguilhem's *The Normal and the Pathological* for an understanding of life as openness to "error." For Reid, this is however based on a failure to recognize that such ontologies of life are already contained in Nietzsche, for whom error was a fully integral dimension of life too. Life is becoming, and truths are created in a perpetual war with error, and Reid here rather looks towards the interpretation proposed by Deleuze as a way to understand biopolitics in an active way.

B) Governing space

Łukasz Stanek's contribution addresses the spatial and architectural implications of Foucault's work, through a discussion of the "scalar organization" of society set against the post-war transformations of urban planning as a biopolitical project. The crisis of Fordism and the welfare state, he suggests, is the crisis of a specific scale, i.e. that of the nation-state, which earlier was able to integrate the scalar levels of country, city, community, and individual house into an organic whole, here exemplified by Polish architect Oskar Hansen's idea of a Continuous Linear System. In a development that began in the later phases of CIAM, in Team X, and in the theories of Aldo van Eyck—and was further intensified in the burgeoning globalization and the deconstruction of the welfare state in the 1970s—scalar levels started to be understood less as bounded entities and more as in-between realms or interstitial spaces, often being understood as sites of political struggle.

Helena Mattsson discusses the corporate takeover of the production of public spaces, and how architectural boundaries between interiority and exteriority, public space and workplace, are increasingly transformed into a pervasive transparency. This, she suggests, has resulted in the creation of "event zones," where production and spectacle come together, also as a means of compensating for the gaps and losses in our understanding of the real processes of production and consumption on a global scale. These are assemblages geared towards the production of a public, a public that, precisely, is seen more as consumers of a spectacle. Such architectural structures should not be understood as disciplinary, but rather as "spaces of security" in Foucault's sense of the term. Instead of regulating everything by clear-cut spatial divisions, these spaces "let things happen." Mattsson argues, however, that they entail new forms of discipline that operate through desire and affect instead of regulation, and in fact can be re-connected to certain aspects of Bentham's Panopticon overlooked by Foucault.

Warren Neidich's contribution looks into how forms of visual and aesthetic experience in general—in the arts, but also in advertising and other forms of communication—can be understood as an exercise of power where the *bios* is increasingly displaced by the *nous*, i. e., a move from *biopolitics* to *noo-* or even *neuorpower* as a mode of power that aspires to directly affect us at the level of neural structures. Drawing on examples ranging from contemporary neuroscience to the works of John Cage, Neidich suggests that art has the capacity to redraw the lines along which the sensible is distributed (in the sense of the "distribution of the sensible" proposed by Jacques Rancière), not just in terms of cultural symbolism, but also on the level of neural maps. In a piece like Cage's *4'33*, Neidich finds a stretching and disruption of time and habits of reception that were undoubtedly difficult or even impossible to receive for its initial audience. These neurological redistributions continue today, but with different means, such as research on an expanded idea of sound and noise, in the context of music.

C) Reconfiguring subjectivity and agency

Cecilia Sjöholm's contribution deals with Foucault and psychoanalysis, and attempts to elaborate a common ground between his later work and Lacan. Drawing on Lacan's first seminar from 1953, *Freud's Papers on Technique*, Sjöholm argues that Foucault's presentation of psychoanalysis as the enterprise of uncovering hidden truths in the subject is mistaken. Lacan in fact abandoned an analysis of the ego in favor of desire, and the question "who am I?" gave way to another question: "who is talking?" Framed as a technique of transference and counter-transference, psychoanalysis concerns the forming rather than the truth of the subject, which is compatible with Foucault's later ideas on self-creation and his quest to revive the ancient care of the self. These convergences are not just accidental resemblance of terminology, Sjöholm argues: both Lacan and Foucault conceived of the formation of the subject as a response to a structure that is always in place, rather than as an adaptation to a norm.

Maurizio Lazzarato investigates Jacques Rancière's claim that Foucault, being occupied with power, never took any interest in political subjectivation. In fact, he suggests, they propose two radically heterogeneous conceptions of political subjectivation: for Rancière, ethics is what neutralizes politics, whereas Foucault's political subjectivation is indistinguishable from a project of *ethopoiesis*, or the formation of the subject. It is a difference of opinion that also comes across in their radically divergent interpretations of Greek democracy.

Lazzarato perceives a logocentric prejudice in Rancière's understanding of democracy. Democracy is, according to Rancière, uniquely based on equality, which in its turn is rooted in the common fact of language: all speech presupposes a mutual understanding and a belonging to a shared community, and political action means to further this possibility and include thereby those that "have no part" in the common. Against this, Lazzarato pits Foucault's understanding of *parrhesia*, a "truth-telling" that introduces paradoxical relations into the formal equality of democracy, as well as a difference of enunciation in the equality of language, and also implies an "ethical differentiation." To take a stand and speak the truth is to take a risk, it pits equals against each other and may lead to hostility and even war; and yet it is the precondition for the production of new forms of subjectivation and singularity, which according to Lazzarato, are downplayed in Rancière's formal understanding of democracy.

Adeena May too looks at Rancière's critique of Foucault, but with the specific aim of readdressing claims made around autism as a form of contemporary subjectivity. When Rancière disputes that politics is reducible to power relations, May argues, this not only means that it is rare, but also that politics must be understood as counterposed to both the "police" and to biopolitical individuation; politics only emerges when the police order and its forms of individuation are disrupted by a different logic, which Rancière calls the logic of "equality." May utilizes Rancière's critique to readdress the question of autism, with respect specifically to the idea of a "neuro-minority," as proposed by the autism rights movement. The question is if these self-defining claims can give rise to political subjectivation, or if they simply reiterate positions within a consensual order. Drawing on Rancière and sociologist Mariam Fraser, May suggests a different approach to such a subjectivation that goes beyond the position of mere resistance, and also allows for a reconsideration of the theoretical tools used to describe such a position.

*

The texts in this collection were originally presented at the conference "The Politics of Life: Michel Foucault and the Biopolitics of Modernity," organized by The Department of Culture and Communication, Södertörn University, and the International Artists Studio Program (IASPIS), in Stockholm, September 3-5, 2009. We would like to thank the staff at IASPIS, The Foundation for Baltic and East European Studies whose generous support was essential in organizing the symposium, the Publication Committee of Södertörn University for making the production of this book possible, and David Payne for his English proofreading.

Stockholm January 2013

Foucault, Politics, and Failure
A Critical Review of Studies of Governmentality

Thomas Lemke

Foucault's work on governmentality, alongside his lectures of 1978 and 1979, delivered at the Collège de France,[1] have inspired many historical investigations as well as studies in the social sciences. The first to further elaborate and develop this "direction for research,"[2] were his fellow researchers: François Ewald, Daniel Defert, Giovanna Procacci, Pasquale Pasquino and Jacques Donzelot, all of whom carried out genealogical investigations into insurance technology, social economy, police science, and the government of the family.[3] Their work mainly focused on the transformations of governmental technologies during the nineteenth century, while French historians like Dominique Séglard, Christian Lazzeri, Dominique Reynié and Michel Senellart used the notion of government to analyze State Reason and early modern arts of government.[4]

---

[1] Michel Foucault, *Security, Territory, Population: Lectures at the Collège de France, 1977-78* (New York: Palgrave, 2007), and The Birth of Biopolitics: Lectures at the Collège de France, 1978-79 (New York: Palgrave, 2008).

[2] Michel Foucault, "*Omnes et Singulatim*: Toward a Critique of Political Reason," in James D. Faubion (ed.), *Power*, trans. R. Hurley et al (New York: The New Press, 2000), 325.

[3] Jacques Donzelot, *L'invention du social: Essai sur le déclin des passions politiques* (Paris: Fayard, 1984); Daniel Defert, "'Popular Life' and Insurance Technology," in Graham Burchell, Colin Gordin, and Peter Miller (eds.), *The Foucault Effect: Studies in Governmentality* (Hemel Hempstead: Harvester Wheatsheaf, 1991); Pasquale Pasquino, "Criminology: The Birth of a Special Knowledge," ibid; François Ewald, *Histoire de l'État providence* (Paris: Grasset, 1996); Giovanna Procacci, *Gouverner la misère: La question sociale en France 1789-1848* (Paris: Seuil, 1993).

[4] See Christian Lazzeri and Dominique Reynié (eds.), *La raison d'état: Politique et rationalité* (Paris: PUF, 1992); Dominique Séglard, "Foucault et le problème du gouvernement," in ibid; Michel Senellart, "Michel Foucault: 'Gouvernementalité' et raison d'Etat," *Penseé Politique* 1 (1993): 276-303; Yves-Charles Zarka (ed.), *Raison et déraison d'Etat* (Paris: PUF, 1994).

Over the last twenty years, however, a new line of reception has taken form in the English-speaking world. While the interest of the French scholars was either genealogical or historical, what has come to be called "Governmentality studies" has mainly addressed contemporary forms of government, focusing, for example, on transformations from welfarism to neo-liberal rationalities and technologies. The publication of the collection *The Foucault Effect: Studies in Governmentality* from 1991 was, in this respect, a significant event. The volume, co-edited by Graham Burchell, Colin Gordon and Peter Miller, presented English translations of Foucault's already published lecture of 1978, entitled "Governmentality,"[5] as well as other important texts. It also made available articles by researchers directly affiliated with Foucault (Defert, Ewald and Donzelot), bringing them together with Anglophones, such as Colin Gordon, Graham Burchell and Ian Hacking.[6] *The Foucault Effect* marked the beginning of a renewed interest in Foucault's work, particularly in Britain, Australia and Canada. In the following years a great number of studies were published, mostly focusing on the rise of neo-liberal forms of government.[7] This interest was not isolated to the Anglo-American context; it applies equally to Scandinavia, Germany, France and in other countries where scholars have sought both to refine and extend Foucault's work on governmentality as an effective means to critically analyze political technologies and rationalities in contemporary societies.[8]

[5] Michel Foucault, "Governmentality," in Burchell et al, *The Foucault Effect*.
[6] As one of the editors, Colin Gordon, frankly admitted nearly 20 years after the publication of the book, the *The Foucault Effect* was "an attempt to construct a plane of consistence between the work of individuals who, in some cases, had never met, and in others were no longer collaborators or desiring to be perceived as such." Jacques Donzelot and Colin Gordon "Governing Liberal Societies: The Foucault Effect in the English-Speaking World," *Foucault Studies* 5 (2008): 48-62, citation at 50. What is more, apart from the editors and Foucault, no contributor to the volume explicitly used the term "governmentality." See Sylvain Meyet, "Les trajectoires d'un texte: 'La gouvernementalité' de Michel Foucault," in Sylvain Meyet, Marie-Cécile Naves, and Thomas Ribemont (eds.), *Travailler avec Foucault: Retours sur le politique* (Paris: L'Harmattan, 2005).
[7] See e.g. Andrew Barry, Thomas Osborne, and Nikolas Rose (eds.), *Foucault and Political Reason: Liberalism, Neo-liberalism and Rationalities of Government* (London: UCL Press, 1996); Mitchell Dean and Barry Hindess (eds.), *Governing Australia: Studies in Contemporary Rationalities of Government* (Cambridge: Cambridge University Press, 1998); Mitchell Dean and Paul Henman, "Governing Society Today: Editors' Introduction," *Alternatives* 29 (2004): 483-94.
[8] See e.g. Ulrich Bröckling, Susanne Krasmann, and Thomas Lemke (eds.), *Gouvernementalität der Gegenwart: Studien zur Ökonomisierung des Sozialen* (Frankfurt am Main: Suhrkamp, 2000), and *Glossar der Gegenwart* (Frankfurt am Main: Suhrkamp, 2004); Lene Koch, "The Government of Genetic Knowledge," in Susanne Lundin and Lynn

This boom in studies of governmentality occurred for theoretical as well as political reasons. During the 1970s and 1980s, many radical intellectuals became increasingly dissatisfied with classical Marxist forms of analysis and critique. Modes of explanation that engaged in some form of economic reductionism—relying on the dogmatic model of base and superstructure, and functionalist concepts of ideology as "false consciousness"— lost a great deal of theoretical credibility. While some scholars tried to combine Marxist concepts with poststructuralist theory, others regarded their interests in cultural forms, subjectivity and discursive processes as serving to give expression to a "post-Marxist" orientation.[9] But the growing reception of the concept of governmentality did not only evolve on a purely theoretical level, it was also linked to a changing political context. In the 1980s and 1990s neoliberal programs and market-driven solutions increasingly replaced Fordist and welfarist modes of government in many countries. These radical transformations called for new theoretical instruments and analytical tools to account for their historical conditions of emergence.

Studies of governmentality have been extremely helpful in illuminating the "soft" or "empowering" mechanisms of power, demonstrating in what ways individuals and social groups are governed by freedom and choice. They have successfully exposed the paradoxes of "controlled autonomy" in neoliberal governmentality and the intimate relationship that exists between the universal call for "self-determination" and quite specific societal expectations and institutional constraints. However, as Jacques Donzelot observes, studies of governmentality may provoke "mixed feelings of pleasure and unease."[10] While, certainly, they have provided promising tools for the analysis of neoliberalism and transformations in (contemporary) statehood, there are also several limitations to be noted—some of which can be traced back to certain ambiguities in Foucault's work.

Åkesson (eds.), *Gene Technology and Economy* (Lund: Nordic Academic Publishers, 2002), and Sylvain Meyet et al, *Travailler avec Foucault*. For overviews of studies of governmentality, see Mitchell Dean, *Governmentality: Power and Rule in Modern Society* (London/Thousand Oaks/New Delhi: Sage, 1999): Thomas Lemke, "Neoliberalismus, Staat und Selbsttechnologien: Ein kritischer Überblick über die governmentality studies," *Politische Vierteljahresschrift* 41(1) (2000): 31-7; Jack Z. Bratich, Jeremy Packer, and Cameron McCarthy (eds.), *Foucault, Cultural Studies, and Governmentality* (Albany: SUNY Press, 2003); Sylvain Meyet, "Les trajectoires d'un texte"; Nikolas Rose, Pat O'Malley, and Mariana Valverde, "Governmentality," in *Annual Review of Law and Social Science* 2 (2006): 83-104.

[9] Rose et al, "Governmentality," 85-89.

[10] Donzelot and Gordon, "Governing Liberal Societies," 53

In this article I will focus on some of the shortcomings and blind spots present in governmentality studies. I will address three problems in particular: first, the idea of a historical succession of sovereignty, discipline and government, prominent in the literature on governmentality; second, some limitations in the analysis of programs and the role of failure in studies of governmentality; and third the question of how politics, materiality and space are conceived in this research perspective.

Sovereignty, discipline, government

Foucault' use of the terms "government" and "governmentality" is marked by inconsistency, tending to change over time.[11] In a very broad sense, government refers to the "conduct of conduct,"[12] and designates rationalities and technologies that seek to guide human beings. Here "governmentality" denotes power relations in general, and Foucault employs the term in order to gain an "analytical grid for these relations of power."[13] In a more specific sense, governmentality refers to a quite distinct form of power. It stands for a historical process closely connected to the emergence of the modern state, the political figure of "population," and the constitution of the economy as a specific domain of reality. This process is characterized by the "pre-eminence over all other types of power—sovereignty, discipline, and so on—of the type of power that we can call 'government.'"[14] In this latter interpretation, Foucault seems to endorse the idea of a continuous shift or historical succession of sovereignty, discipline and government.

Building on this idea, there has been a tendency in the governmentality literature to use the notion of governmentality as a historical meta-narrative that leads from state reason, via classical liberalism and the welfare state, to contemporary neoliberal forms of government.[15] Government has been seen

[11] See Thomas Lemke, *Eine Kritik der politischen Vernunft: Foucaults Analyse der modernen Gouvernementalität* (Hamburg/Berlin: Argument, 1997), 197-8; Mitchell Dean, *Governmentality*, 16; Michel Senellart, "Course Context," in Michel Foucault, *Security, Territory, Population*, 386-391
[12] Michel Foucault, "The Subject and Power," in James D. Faubion (ed.), *Power*, trans. R. Hurley et al (New York: The New Press, 2000), 341.
[13] Michel Foucault, *The Birth of Biopolitics*, 186
[14] Michel Foucault, *Security, Territory, Population*, 108
[15] See Thomas Osborne, "Techniken und Subjekte: Von den 'Governmentality Studies' zu den 'Studies of Governmentality,'" in Wolfgang Pircher and Ramón Reichert (eds.),

as some superior form of rule that unfolds in Western modernity, suggesting the displacement or marginalization of sovereign law and disciplinary technologies.[16] Studies of governmentality, then, have often assumed a continuous rationalization of forms of government, while discipline and sovereignty have been conceived as accidental, auxiliary or residual, modes. According to this line of interpretation, discipline and sovereignty will sooner or later be replaced by governmental technologies, which are taken to be more "economic." Many authors have held that actuarial techniques of power are reducing social conflict and provoking less resistance to social regulation, while, at the same time, increasing the effective government of populations by improving the productivity of labor, the health of the population, and so on. By contrast, discipline and sovereignty have often been regarded as archaic and redundant technologies of power, so that authors employing a governmentality perspective came to note a fundamental change from sovereign mechanisms and disciplinary technologies to a "risk society,"[17] or a "post-disciplinary order."[18]

This teleological reading of governmentality presupposes a stable "economy of power" and regards "efficiency" as some kind of absolute and universal yardstick, making it possible thereby to compare and rank hierarchically different technologies of power, in terms of certain goals they are already expected to achieve. The problem is that such a conception of technology tends toward a certain form of idealism and therefore is completely inadequate as a way of understanding how technologies change over time and how they interact with one another. As Pat O'Malley rightly concludes, technologies like actuarism vary in accordance with different historical and spatial contexts—as well as in light of specific articulations with other technologies and programs. The displacement of one technology of power by another cannot be measured in abstract terms or by an immanent logic of gradual improvement and progress. Rather, the process is

---

Governmentality Studies: Analysen liberal-demokratischer Gesellschaften im Anschluss an Michel Foucault (Münster: Lit-Verlag, 2004).
[16] See Tania Murray Li, *The Will to Improve: Governmentality, Development, and the Practice of Politics* (Durham and London: Duke University Press, 2007), 12.
[17] See Jonathan Simon, "The Emergence of a Risk Society: Insurance, Law, and the State," *Socialist Review* 95 (1987): 61-89.
[18] See Robert Castel, "From Dangerousness to Risk," in Graham Burchell et al, *The Foucault Effect*.

very much political; the result of struggles and conflicts, of altered compromises and new alliances.[19]

There is a second problem with the idea of a continuous displacement and rationalization of technologies of rule. Studies of governmentality tend to emphasize the "productive" side of power at the expense of the investigation of "repressive" and authoritarian mechanisms. At the center of the analytical interest are governmental technologies that operate not by exercising violence and constraint but by effecting "powers of freedom."[20] Such works often ignore or underestimate the role of violent and "irrational" forms of politics, e.g. the mobilization of fear or seemingly "uneconomic" populist discourses. By adhering to a rather abstract concept of rationality, studies of governmentality have tended to neglect the political significance of expressive and emotional factors in favor of conscious calculations and elaborated concepts.[21] Especially since 9/11, the intimate relationship between governmentality and sovereignty, between neoliberalism and discipline, freedom and violence, can no longer be ignored.

The thesis of a continuous rationalization of power is not only wrong because it obscures the enduring significance of repression and violence in contemporary forms of rule. More fundamentally, it ignores the internal relationship and co-determination between "rational" and "irrational" elements, freedom and authoritarianism, that characterize (neo-)liberal government. Mariana Valverde has, for example, argued that the constitution of the liberal subject not only necessitates a permanent work of moralization and disciplination of the self; it also makes possible the governing of "backward" or "primitive" races, classes or sexes in order to bring them up to the level of autonomous liberal subjects—with the use of disciplinary or "despotic" techniques.[22] In the same vein, Barry Hindess has

---

[19] See Pat O'Malley, "Risk and Responsibility," in Barry et al, *Foucault and Political Reason*, 192-198.
[20] See Nikolas Rose, *Powers of Freedom: Reframing Political Thought* (Cambridge: Cambridge University Press, 1999). David Garland has stressed that the governmentality literature tends not to distinguish adequately between the concept of agency and the concept of freedom. They are often conflated, but it is important to insist on their difference: "The truth is that the exercise of governmental power, and particularly neoliberal techniques of government, rely upon, and stimulate, *agency* while simultaneously reconfiguring (rather than removing) the *constraints* upon the freedom of choice of the agent." Garland, "'Governmentality'" and the Problem of Crime: Foucault, Criminology, Sociology," *Theoretical Criminology* 1(2) (1997): 199-204, cit. at 197, emphasis in original.
[21] See David Garland, "'Governmentality' and the Problem of Crime."
[22] Mariana Valverde, "'Despotism' and Ethical Liberal Governance," *Economy & Society* 25(3) (1996): 357-72.

insisted that sovereign and authoritarian measures cannot be regarded as auxiliary or secondary within liberal rationalities, since they are actually constitutive of them; liberty and domination are two sides of the same coin in liberal governmentality. Rather than representing a denial of the commitment to liberty, "the resort to authoritarian rule in certain cases is a necessary consequence of the liberal understanding of that commitment."[23] In this light the persistence of illiberal practices is not an accidental side effect, a matter of hypocrisy or a logical contradiction; rather, (neo-)liberal rationalities are characterized by a specific articulation of autonomous subjectivation and disciplinary subjection, freedom and domination.

The teleological interpretation also runs against Foucault's insistence on the "overlappings, interactions and echoes"[24] of different power technologies.[25] Foucault regularly stressed that sovereignty, discipline and government do not succeed or substitute one another but are mutually supportive: "We should not see things as the replacement of a society of discipline by a society, say of government. In fact we have a triangle: sovereignty, discipline, and governmental management."[26] The focus then is not on homogeneous and abstract types of rule but on assemblages, amalgams and hybrids—put otherwise, on the heterogeneous and concrete ways different technologies interact.[27] This considered, an analytics of government should be attentive to the co-existence, complementarity and interference of different technologies of rule.

Consequently, studies of governmentality not only have to assume a plurality of rationalities and technologies, they also have to understand them to be plural, messy and contradictory.

---

[23] Barry Hindess, "The Liberal Government of Unfreedom," *Alternatives* 26 (2001): 93-111, cit. at 94. See also Mitchell Dean, "Liberal Government and Authoritarianism," *Economy and Society* 31(1) (2002): 37-61, and *Governing Societies: Political Perspectives on Domestic and International Rule* (Maidenhead, Berkshire: Open University Press, 2007), 108-129.

[24] Michel Foucault, *The History of Sexuality, Vol. 1: An Introduction* (New York: Vintage 1980), 149.

[25] Pat O'Malley, "Risk and Responsibility," 192.

[26] Michel Foucault, *Security, Territory, Population*, 107

[27] See Stephen Gill, "The Global Panopticon? The Neoliberal State, Economic Life, and Democratic Surveillance," *Alternatives* 20(2) (1995): 1-49; David Scott, "Colonial Governmentality," in Jonathan X. Inda (ed.), *Anthropologies of Modernity: Foucault, Governmentality, and Life Politics* (Malden, MA/Oxford: Blackwell 2005); Tania Murray Li, *The Will to Improve*, 12-17; Brian C. Singer and Lorna Weir, "Sovereignty, Governance and the Political: The Problematic of Foucault," *Thesis Eleven* 94 (2008): 49-71; Michael Dillon and Julian Reid, *The Liberal Way of War: Killing to Make Life Live* (London/New York: Routledge, 2009).

Programs, strategies, and failure

Studies of governmentality distance themselves from realist sociology and from "sociologies of rule" that study the ways in which rule is actually accomplished. By contrast, work on governmentality focuses on the projects and programs of government, on rationalities and technologies rather than on their outcomes and effects.[28] This self-understanding parallels Foucault's explicit interest in the lectures on governmentality in investigating "the art of governing, that is to say, the reasoned way of governing best and, at the same time, reflection on the best possible way of governing' or 'government's consciousness of itself."[29] Taking up this line of investigation, studies of governmentality have analyzed "mentalities of rule."[30] This does not mean that the research has focused on ideal types and normative inter-pellations. Rather, studies of governmentality have examined governmental programs as empirical facts, insofar as they shape and transform the real by providing specific forms of representing and intervening in it.

While it has rarely been disputed that studies of governmentality focus on programs, it is the way such programs have been analyzed that has given rise to a number of problems. First, some authors have tended to treat programs as closed and coherent entities, as achievements and accomplishments rather than as projects and endeavors. They have often explicated in what ways programs have successfully obscured political alternatives, obstructing resistance and opposition. Governmental programs were often depicted as totalizing and powerful, while contestation remains residual and

[28] See Colin Gordon, "Governmental Rationality: An Introduction," in Burchell et al, *The Foucault Effect*; Mariana Valverde, "'Despotism' and Ethical Liberal Governance," *Economy & Society* 25(3) (1996): 357-72; Mitchell Dean, "Questions of Method," in Irving Velody and Robin Williams (eds.), *The Politics of Constructionism* (London: Sage, 1998); Nikolas Rose, *Powers of Freedom*, 19-20.
[29] Michel Foucault, *The Birth of Biopolitics*, 2. Foucault explained the distinctiveness of this kind of analysis in a response to French historians: "You say to me: nothing happens as laid down in these 'programs'; they are not more than dreams, utopias, a sort of imaginary production that you aren't entitled to substitute for reality. [...] To this I would reply: if I had wanted to describe 'real life' in the prisons, I indeed wouldn't have gone to Bentham. But the fact that this real life isn't the same thing as the theoreticians' schemes doesn't entail that these schemes are therefore utopian, imaginary, and so on. [...] [T]hese programs induce a whole series of effects in the real (which isn't of course the same as saying that they take the place of the real): they crystallize into institutions, they inform individual behavior, they act as grids for the perception and evaluation of things." Michel Foucault, *Fearless Speech*, ed. Joseph Pearson (New York: Semiotext(e), 2001), 232.
[30] Peter Miller and Nikolas Rose, *Governing the Present: Administering Economic, Social and Personal Life* (Cambridge: Polity, 2008), 20; 24; Mitchell Dean, *Governmentality*, 16.

marginal. However, opposition and struggles do not only take place in an interval "between" programs and their "realization"; they are not limited to some kind of negative energy or obstructive capacity. Rather than "distorting" the "original" plans, they are instead always-already part of them, actively contributing to "compromises," "fissures" and "incoherencies" constitutive of such programs. Thus, an analytics of government must take into account the "breaks" or "gaps" interior to programs—viewing them not as signs of their failure but as the very condition of their existence.[31]

There is a second tendency in the governmentality literature that contrasts and complements the first. Many authors have stressed the importance of "failure," regarding government as a permanently failing operation.[32] Failure stands here for the collision between program and reality. While this reading rightly subverts the idea of a closed and coherent program or idealized scheme—in the stress that it places on the fragility and the dynamic aspect of government—the focus on failure is nonetheless somewhat ambivalent. As Pat O'Malley remarks, failure is "not an intrinsic property of an event so much as it is a property of a program. To think in terms of failure puts the emphasis on the status of the collision from the programmer's viewpoint, and consequently reduces resistance to a negative externality."[33] While "failure" points to the incompleteness and contingencies of governmental programs, it inadvertently reduces the role of opposition, struggle and conflict to that of obstruction and refusal. For many studies of governmentality contestation is not part of the programs— and its role remains purely negative and limited to resistance. As a consequence, the constructive (and not only obstructive) role of struggles, and the ways in which opposition and rule interact, tend not to be analyzed.[34]

---

[31] See Lorna Weir, "Recent Developments in the Government of Pregnancy," *Economy & Society* 25(3) (1996): 373-92; Pat O'Malley, "Indigenous Governance," *Economy & Society* 25(3) (1996): 310-26; Thomas Lemke, "Neoliberalismus, Staat und Selbsttechnologien"; Tania Murray Li, *The Will to Improve.*

[32] See Alan Hunt and Gary Wickham, *Foucault and Law: Towards a Sociology of Law as Governance* (London: Pluto Press 1994); Jeff Malpas and Gary Wickham, "Governance and Failure: On the Limits of Sociology," *Australian and New Zealand Journal of Sociology* 31(3) (1995): 37-50; Peter Miller and Nikolas Rose (2008) *Governing the Present*, 35

[33] Pat O'Malley, "Indigenous Governance," 311.

[34] Andrew Barry notes that the notion of "resistance" provides only an impoverished idea of the dynamics of contestation and opposition: "Following Foucault's own work, there has been a lack of interest in the analysis of study of political conflict, and a tendency to resort, in the absence of any developed account, to the notion of 'resistance' to understand such conflicts." Andrew Barry, *Political Machines: Governing a Technological Society* (London: Athlone Press, 2001), 199.

Thus it seems the focus on failure is insufficient. To contrast rationalities and technologies of government does not trace any clash between program and reality, the confrontation of the world of discourse and a field of practices. The relations between rationalities and technologies, programs and institutions, are much more complex than a simple application or transfer. The difference between the envisioned aims of a program and its actual effects does not refer to the distance between the purity of the program and the messy reality, but, rather, to different layers of reality.

To capture this dynamic relationship, it might be useful to take into account Foucault's insistence on the strategic character of government. In contrast to many studies of governmentality, Foucault not only shows that government "fails" or how it gives rise to unintended effects. Moreover he takes into account that actors respond to changing outcomes, calculating and capitalizing upon them and integrating them into their future conduct.[35] Let me illustrate this through an example Foucault provides in *Discipline and Punish*, namely the "failure" of the prison system, which produced delinquency as an unintended effect. In his genealogy of the prison Foucault does not confront program and reality, nor does he frame the problem in terms of functionality. The institutionalization of the prison in the nineteenth century produced

> an entirely unforeseen effect which had nothing to do with any kind of strategic ruse on the part of some meta- or trans-historic subject conceiving and willing it. This effect was the constitution of a delinquent milieu [...]. The prison operated as a process of filtering, concentrating, professionalizing and circumscribing a criminal milieu. From about the 1830s onward, one finds an immediate re-utilization of this unintended, negative effect within a new strategy which came in some sense to occupy this empty space, or transform the negative into a positive. The delinquent milieu came to be re-utilized for diverse political and economic ends, such as the extraction of profit from pleasure through the organization of prostitution. This is what I call the strategic completion (*remplissement*) of the apparatus.[36]

Emphasizing the strategic dimension of government allows the focus to be placed on the conflicts and contestations advanced against the very techno-

---

[35] Tania Murray Li, *The Will to Improve*, 287.
[36] Michel Foucault, "The Confession of the Flesh," in Michel Foucault, *Power/Knowledge: Selected Interviews and Other Writings*, ed. Colin Gordon (New York: Pantheon Books, 1980), 195-96.

logies and rationalities constituting governmental practices. Political struggles cannot be confined to the expression of a contradictory logic or an antagonistic relation; they have their own dynamics, temporalities and techniques.[37] With due focus on the "parasitic relationship"[38] of governments—honing in specifically upon their "failures" and "shortcomings"— what becomes possible is the, circumventing of any functionalist bias. If contestation is limited to the refusal of programs, then the following question arises: what exactly does "failure" mean? Since the criteria of judging both failure and success are an integral part of rationalities, they cannot be regarded as external yardsticks. In fact, the "success" of a program is no guarantee of its continuation, since success might eventually abolish the material foundations or preconditions for a given program, making it redundant thereby. Conversely, the putative "failure" of a program could mean its "success," since it might give rise to "strategic reinvestment." Put differently: a program might work "well" because it does not work at all or only works "badly," for example, by creating the very problems it is supposedly there to react to. Therefore, the "failure" of the prison as a means to combat criminality might possibly help to account for its *"raison d'être."*[39]

Politics, materiality, and space

Jacques Donzelot has pointed out a tendency in the governmentality literature to treat governmental regimes as things that "are always analyzed at their 'technical' level, never in terms of a political criterion or in terms of value."[40] According to Donzelot, the rather neutral rhetoric deployed in governmentality studies is the result of a dual process: on the one hand, the deliberate focus on the programmatic and technical aspects of government and, on the other, an insight into the problems associated with reductionist and simplistic forms of analyzing and criticizing neoliberalism. While Foucault considered himself a political intellectual—actively engaged with the

---

[37] Andrew Barry, *Political Machines*, 6. According to Foucault, power relations and "strategies of struggle" are characterized by an agonistic relationship: "a relationship that is at the same time mutual incitement and struggle; less of a face-to-face confrontation that paralyzes both sides than a permanent provocation." Foucault, "The Subject and Power," 342.
[38] Tania Murray Li, *The Will to Improve*, 1.
[39] See also Albert. O. Hirschman, *The Passions and the Interests: Political Arguments for Capitalism Before its Triumph* (Princeton, NJ: Princeton University Press, 1977).
[40] Jacques Donzelot and Colin Gordon, "Governing Liberal Societies," 54.

social movements of his time—the effects of governmental regimes are rarely assessed in Foucault's legacy. As Tania Murray Li has observed, studies of governmentality "tend to be anemic on the practice of politics."[41]

In fact, the critical distance governmentality studies places between itself and forms of social critique—which it labels as reductive—has often resulted in an impasse, serving to limit its own critical engagement. This distancing from critique shows itself when such studies routinely remain at the descriptive level of analyzing rationalities and technologies. With an intention of going beyond "negative" forms of critique, either in the form of condemning or denouncing social and political reality, some authors have surmised critique *per se* to be solely a negative enterprise. The outcome has been a "rhetorical strategy that poses genealogical work over and against criticism."[42] Following this impetus, some authors have explicitly stated that they do not wish "to provide a 'critique' of various liberal and neoliberal problematizations of government" by drawing "a balance sheet of their shortcomings or to propose alternatives."[43] Indeed, the question as to what governmentality studies may offer in the way of a critique of contemporary societies is one for which no single response has been proffered; varying answers have been articulated therefore by a broad range of individual authors. While some seek to redefine and combine governmentality and neo-Marxist concepts,[44] others appear to locate themselves explicitly within a post-Marxist tradition.[45]

[41] Tania Murray Li, *The Will to Improve*, 26; see also Pat O'Malley, Lorna Weir, and Clifford Shearing, "Governmentality, Criticism, Politics," *Economy & Society* 26(4) (1997): 507-508. In extreme cases, studies of governmentality might even contribute to an affirmative reading of governmental rationalities. The most prominent example of this is the trajectory of François Ewald, who was one of Foucault's fellow researchers and undertook a remarkable genealogy of social insurance; see Ewald, *Histoire de l'État providence*. Today, he is a leading representative of the national employers' organization and celebrates the ontology of risk and the virtues of enterprise. See Jacques Donzelot and Colin Gordon, "Governing Liberal Societies," 53; 55; see also Maurizio Lazzarato, "Le gouvernement par l'individualisation," *Multitudes* 2 (2001): 153-61.
[42] Pat O'Malley, Lorna Weir, and Clifford Shearing, "Governmentality, Criticism, Politics," 504.
[43] Andrew Barry, Thomas Osborne, and Nikolas Rose, "Liberalism, Neo-Liberalism and Governmentality: Introduction," *Economy & Society* 22(3) (1993): 266.
[44] See Frank Pearce and Steve Tombs, "Hegemony, Risk and Governance: 'Social Regulation' and the American Chemical Industry," *Economy & Society* 25(3) (1996): 428-54; Tania Murray Li, *The Will to Improve*; Bob Jessop, "From Micro-Powers to Governmentality: Foucault's Work on Statehood, State Formation, Statecraft and State Power," *Political Geography* 26 (2007): 34-40.
[45] See e.g. Nikolas Rose, "Government, Authority and Expertise in Advanced Liberalism," *Economy & Society* 22(3) (1993): 283-99; Peter Miller and Nikolas Rose, *Governing the Present*, 2-4

There seems, however, to be a misunderstanding regarding the role of politics in the literature on governmentality.[46] While it is certainly right to stress the distinctiveness of studies of governmentality, this should not result in a reluctance to evaluate the effects of governmental regimes. It is perfectly possible to emphasize that the study of governmental rationalities and social history are different kinds of inquiry, requiring specific tools of analysis, without one being privileged over the other or one separated from another. Rather than confronting them, it seems more fruitful to investigate their co-production and dynamic interactions and therefore to examine empirically how programs are constituted, transformed and contested.[47]

Ironically, studies of governmentality not only suffer from being too focused on the technical, often they do not take the technical seriously enough. As Andrew Barry has remarked, there has been "little attempt, with a few exceptions, to integrate some of the insights of Foucauldian approaches to the study of government with the work of the anthropologists, sociologists and historians of science and technology."[48] Most authors using the concept of governmentality tend to reproduce a rather classical bias already present in the social sciences. They take the realm of the social as self-evident and "natural," accordingly concentrating their investigations on the activity of humans at the same time as regarding technological devices as inert and passive. Science and technology studies (STS) have, in the last thirty years, shown the limits of this perspective, and are attentive to how socio-material practices shape and transform reality, giving rise to a multitude of actors and resulting in different "ontological politics."[49] This area of research points to the hybridity of actors and networks, focusing on the arrangements that assemble human and non-human actors, living beings and technological artifacts, material bodies and symbolic structures.[50]

[46] See also Barry Hindess, "Politics and Governmentality," *Economy & Society* 26(2) (1997): 257-72.
[47] Tania Murray Li, *The Will to Improve*, 27
[48] Andrew Barry, *Political Machines*, 199-200
[49] See Annemarie Mol, "Ontological Politics: A Word and Some Questions," in John Law and John Hassard (eds.), *Actor Network and After* (Oxford: Blackwell, 1999).
[50] See Michel Callon, "Some Elements of a Sociology of Translation: Domestication of the Scallops and the Fishermen of St. Brieue Bay," in John Law (ed.), *Power, Action and Belief* (London: Routledge, 1986); Donna Haraway, *Simians, Cyborgs, and Women: The Reinvention of Nature* (London: Routledge, 1991); John Law (ed.), *A Sociology of Monsters: Essays on Power, Technology and Domination* (London: Routledge, 1991), and Bruno Latour, "Technology is Society Made Durable," in ibid. While there is a general tendency in studies of governmentality to neglect the governmental dimension of socio-

The principal focus of the governmentality literature on the governing of humans (and the simultaneous analytical lack of interest in technical artifacts and non-human nature) as well as the "social" as the unquestioned plane of reference goes back to Foucault's work, where government is mostly understood as the guidance of human conduct.[51] However, once again a more fruitful reading is possible. In the 1978 lecture series from the Collège de France, Foucault refers to a definition of government provided by Guillaume de la Perrière in an early modern tract on the art of government. Here, government is conceived of as the "right disposition of things."It is concerned with a "complex of men and things": "men in their relationships, bonds, and complex involvement with things like wealth, resources, means of subsistence, the territory with its borders, qualities, climate, dryness, fertility, and so on."[52] From this perspective, government not only focuses on governing humans and the relations that exist between humans. It also refers to a more comprehensive reality that includes the material environment and the specific arrangements and technical networks that relate the human and the non-human. This conceptual shift not only makes it possible to extend the territory of government, multiplying the elements and the relations it consists of; it also initiates a reflexive perspective that takes into account the diverse ways in which the boundaries between the human and the non-human world are negotiated, enacted and stabilized. Furthermore, this theoretical stance makes it possible to analyze the sharp distinction drawn between, on the one hand, the natural and technical from the social on the other, as itself a distinctive instrument and effect of governmental rationalities and technologies.[53]

The fact that the boundaries between bodies, collectives and institutions are permanently undermined, reconfigured and transformed by socio-

---

technical arrangements, there are some important links to the work of STS scholars, especially to actor network theory and Callon's and Latour's idea of a sociology of translation; see Nikolas Rose, *Powers of Freedom*, 49; Peter Miller and Nikolas Rose, *Governing the Present*, 33-4. Furthermore, there exist some innovative projects to combine science and technology studies and an analytics of government: see e.g. Andrew Barry, *Political Machines*; Lene Koch and Mette Nordahl Svendsen, "Providing Solutions – Defining Problems: The Imperative of Disease Prevention in Genetic Counselling," *Social Science and Medicine* 60 (2005): 823-32; Kristin Asdal, "On Politics and the Little Tools of Democracy: A Down-to-Earth Approach," *Distinktion: Scandinavian Journal of Social Theory* 16 (2008): 11-26.
[51] See e.g. *The Birth of Biopolitics*, 2.
[52] *Security, Territory, Population*, 96
[53] See Bruno Latour, *We Have Never Been Modern* (Cambridge, Mass.: Harvard University Press, 1993).

technical arrangements points to another shortcoming of governmentality studies. While this theoretical perspective has been extremely helpful in displacing the idea of the state as the natural and coherent center of power so as to study the plural and heterogeneous character of governmental rationalities and technologies, it is mostly the territorially sovereign nation state that serves as the implicit or explicit frame of reference in the govern-mentality literature. There is rarely any consideration given to how trans-formations of government on a national level are linked up with inter-national developments or to how the appearance of new actors on a global or European scale is paralleled by a shift of the competences of the nation state.[54] The limits of this approach make it difficult to investigate new forms of government, indicated by the increasing significance of international, supranational and transnational organizations like the UN, IMF and World Bank. Furthermore, the approach does not appear to account for the new role of transnational alliances in Nongovernmental Organizations. As James Ferguson and Akhil Gupta rightly stress, it is necessary to extend an analytics of government to include modes of government constituted on a transnational and global scale. They criticize the way in which

> institutions of global governance such as the IMF and the WTO are commonly seen as being simply "above" national states, much as states were discussed vis-à-vis the grassroots. Similarly, the "global" is often spoken of as if it were simply a superordinate scalar level that encompasses nation-states just as nation-states were conceptualized to encompass regions, towns, and villages.[55]

Since the turn of the century, recent discussions surrounding "transnational" or "global governmentality" show scholars to be already rethinking and questioning spatial as well as scalar framings of sovereign states—too often taken for granted in the literature on governmentality.[56]

---

[54] For a notable exception to this general tendency, see Andrew Barry, "The European Community and European Government: Harmonization, Mobility and Space," *Economy & Society* 22(3) (1993): 314-26, and *Political Machines*.

[55] James Ferguson and Akhil Gupta, "Spatializing States: Toward an Ethnography of Neoliberal Governmentality," *American Ethnologist* 29(4) (2002): 981-1002, cit. at 990.

[56] See Wendy Larner and William Walters, *Global Governmentality: Governing International Spaces* (London: Routledge, 2004), and "Globalization as Governmentality," *Alternatives* 29 (2004): 495-514; Richard Warren Perr and Bill Maurer (eds.), *Globalization Under Construction: Governmentality, Law, and Identity* (Minneapolis: University of Minneapolis Press, 2003); Aihwa Ong and Stephen J. Collier (eds.), *Global Assemblages: Technology, Politics, and Ethics as Anthropological Problems* (Malden, MA: Blackwell, 2004); Stuart Elden, "Governmentality, Calculation, Territory," *Environment*

However, there is another aspect to this problem. Not only has the nation state been the privileged reference for studies of governmentality, the analysis has also focused on very specific exemplars: Western liberal democracies. Until recently, studies of governmentality were often informed by a "eurocentrism"[57] ignoring non-Western as well as non-liberal contexts. As Gary Sigley rightly notes, what is at stake here is not a simple "extension" of the governmentality perspective to an as yet neglected object, extending its "application" to a different area of research; rather, "we must accept the possibility that it is not only the perception of the foreign that will be altered but also the original 'theory' itself."[58] Empirical work on non-Western governmental regimes might produce insights and effects that fall back on the use of the concept of governmentality and the way in which studies of governmentality are conducted.

Conclusion

Many of the deficits and blind spots described above have been discussed for a long time in the literature on governmentality.[59] This internal discussion has given rise to a critical self-evaluation of this research perspective, so as to correct these conceptual limitations and analytical shortcomings. However, some important problems surrounding this research perspective do not originate in either ambiguities or "failures" but are rather the outcomes of its success. There is a kind of paradox here. Studies of governmentality have received considerable attention among scholars, since they possess a high degree of diagnostic potential for a critical analysis of the present; but it is exactly the immense attention they receive that seems to undermine their analytical and critical potential. As Nikolas Rose, Pat

---

and Planning D: Society and Space 25(4) (2007): 562-80. On Foucault's account of the organization and politics of space, see Jeremy W. Crampton, and Elden Stuard (eds.), *Space, Knowledge and Power: Foucault and Geography* (Aldershot: Ashgate, 2007), and Stuart Elden, "Governmentality, Calculation, Territory."

[57] James Ferguson and Akhil Gupta, "Spatializing States," 998.

[58] Gary Sigley, "Chinese Governmentalities: Government, Governance and the Socialist Market Economy," *Economy & Society* 35(4) (2006): 487-508, cit. at 488.

[59] See Pat O'Malley, Lorna Weir, and Clifford Shearing, "Governmentality, Criticism, Politics"; Thomas Lemke, "Neoliberalismus, Staat und Selbsttechnologien"; Nikolas Rose, Pat O'Malley, and Mariana Valverde, "Governmentality"; Ulrich Bröckling, Susanne Krasmann, and Thomas Lemke, "From Foucault's Lectures at the Collège de France to Studies of Governmentality: An Introduction," in Ulrich Bröckling, Susanne Krasmann, and Thomas Lemke (eds.), *Governmentality: Current Issues and Future Challenges* (London: Routledge, 2011).

O'Malley and Marianna Valverde put it: "If one of the attractions of governmentality has been its capacity to render neoliberalism visible in new ways
[...], ironically in certain respects this also has become a handicap.
Although some writers have made it clear that neoliberalism is a highly
specific rationality [...], a marked tendency has been to regard it as a more
or less constant master category that can be used both to understand and to
explain all manner of political programs across a wide variety of settings."[60]

Two problems merit special attention. The first is a tendency to
canonize, systematize, and normalize this theoretical perspective in a manner that is to its own detriment. Studies of governmentality are sometimes
regarded as a distinctive "theory" or a specific "school." This description is
problematic insofar as it suggests a level of coherence and elaboration that
the literature on governmentality actually lacks. Authors writing under the
rubric of governmentality have followed different theoretical paths, and
they have chosen a large variety of empirical objects and addressed highly
diverse questions. Not that this "lack" of coherence is itself a problem that
should be resolved in the future, but a quite deliberate stance and, indeed, a
specific strength. There is no single theoretical program or general approach, and there can be no such thing, since governmentality is not a
model or framework of explication but a distinctive critical perspective and
a style of thought. It offers conceptual instruments that point to the "costs"
of contemporary forms of government while providing a basis for the
invention of new practices and modes of thinking.

The second problem concerns repetition and the rise of a "governmentality industry."[61] Many studies observe the same strategies or technologies for an ever growing variety of objects and areas: indirect forms of
government like empowerment or activation instead of direct intervention;
a focus on responsibilization and risk management; development of entrepreneurial modes of action and organization, and so on. The reader already
seems to know in advance what he or she is going to read. As a result, any
surprising insights derived from the empirical data and material are
effectively ruled out. This theoretical trivialization is paralleled by a systematic overvaluation of the concept. While Foucault always formed his
analytical instruments in relation to the historical objects he was concretely
studying (madness, delinquency, sexuality etc.) without providing a "general theory," governmentality has become in some studies a kind of meta-

---

[60] Nikolas Rose, Pat O'Malley, and Mariana Valverde, "Governmentality," 97.
[61] See Thomas Osborne, "Techniken und Subjekte."

narrative to be used for any object of investigation—without it being in need of correction or further development.

While it is certainly necessary to counter tendencies to "canonize" or "generalize" this style of thought, it is also foreseeable that the interests and the focus of the reception will be altered now that Foucault's lectures of 1978 and 1979—along with many of his shorter texts—have been translated into many different languages. New lines of interpretation will invariably develop, addressing different problems and offering new readings of "governmentality." However, it will be important to overcome the theoretical isolationism that has so far prevailed in the governmentality literature. For such isolationism generates the problem that only a few or marginal references in this literature exist to other important forms of contemporary theory, like for example science and technology studies, post-colonial theory and gender studies. Taking up these theoretical traditions, so as to reflect upon governmentality, does not have to result in some kind of general theory, nor does it mean harmonizing different theoretical accounts by ignoring the tensions between them. Rather, it might prove useful to recall that the "tool box" can be used in various ways, depending on the different objects being studied and the objectives being pursued. The productivity of governmentality studies and its critical potential rests on whether or not it is possible to integrate innovative concepts and ideas and concomitantly to open up new research methods and lines of questioning.

Neoliberalism and Biopolitical Governmentality

Johanna Oksala

The factual, empirical account of the rise of neoliberal hegemony is fairly uncontested. The collapse of the Bretton Woods system in 1971 meant that in a floating currency system it was no longer possible to control capital flows or financial markets. The years 1978–1980 represented a further turning point: Margaret Thatcher and Ronald Reagan were elected, Paul Volcker took command at the US Federal Reserve and Deng Xiaoping took decisive steps towards the liberalization of China's economy. These economic and political decisions taken by key actors took the world economy in a new direction. The goal of a pervasive welfare state with the objective of full employment was systematically replaced with the objective of creating an institutional framework that supported free domestic markets, free international trade and individual entrepreneurial conduct. Even though the actual process of implementing these objectives has varied widely in different parts of the world and has in many countries been only partial, on the level of historical and economic facts it is possible to identify a worldwide neoliberal turn in the 1970s.[1]

On the level of historical ontology—the level conditioning our thought and experience of the world—the spectacular rise of neoliberalism is harder to understand and to account for. Rather than being the achievement of a few key actors, it was rooted in much deeper structural and systemic changes in our conception of the political and the practices of governing. It is my contention that in order to engage in any kind of critical evaluation of neoliberalism it is important to study it on this level, too. It is to be understood

---

[1] For some recent empirical accounts of the spread of neoliberalism, see e.g. Alfredo Saad-Filho and Deborah Johnston (eds.), *Neoliberalism: A Critical Reader* (London: Pluto Press, 2004); David Harvey, *A Brief History of Neoliberalism* (Oxford: Oxford University Press, 2005); Naomi Klein, *The Shock Doctrine: Rise of Disaster Capitalism* (London: Allen Lane, 2007).

not merely as an economic doctrine, but also as a comprehensive framework for understanding ourselves and the political reality we live in today.

My aim in this paper is to argue that a Foucauldian ontology of the present provides a valuable and original set of tools for such a philosophical critique of neoliberalism. I will show that the political ontology of neo-liberalism can be effectively explicated along the three axes of power, know-ledge and subjectivity, which Foucault considered central to any critical inquiry into our present. Specifically, I will focus on his lectures on neo-liberalism, *The Birth of Biopolitics*, delivered at the College de France in 1978–79.[2] These lectures analyze the neoliberal program in its two principal forms: the initial German form was represented by the proponents of the Freiburg School of economics such as Walter Eucken and Wilhelm Röpke, also called "Ordoliberals," coined from the journal *Ordo*. It was strongly linked to the critique of Nazism and, after the War, to post-war reconstruction. The other, American form, was the neoliberalism of the Chicago School, which, while deriving from the former, was in some respects more radical.[3]

Diverging from interpretations that treat these lectures as economic or social history, I want to emphasize their philosophical character. As Foucault describes his objective in the first lecture, his interest is in the construction of reality: the focus in his research is to understand the coupling of a set of practices with a regime of truth in order to follow the effects of its inscription on reality (BB, 19). His philosophical claim, in essence, is that neoliberalism functions as an apparatus of knowledge and power: it constructs a particular kind of social and political reality. We have come to understand the world around us in a distinctive way through the matrix of neoliberalism, and this framework delimits our political ration-ality as well as our implicit self-understanding.

My argument proceeds in three stages, following the three axes of knowledge, power and subjectivity. In the first part I claim that neoliberal-ism can be viewed as an extreme form of the liberal regime of truth regulating our current governmentality. Secondly, I show how neoliberalism is compa-

[2] *The Birth of Biopolitics: Lectures at the Collége de France, 1978–1979*, ed. Michel Senellart, trans. Graham Burchell (Basingstoke: Palgrave Macmillan, 2008). All references to these lectures are designated in the text as BB.
[3] There are a number of connections between the two: they share the same enemy—a state-controlled economy—and a series of persons, theories and books passed between them. Yet, they also have their own distinctive features. Foucault argued that the Chicago School was more radical in its expansion of the economic to the social, ultimately eliding the difference between them. See e.g. BB, 323.

NEOLIBERALISM AND BIOPOLITICAL GOVERNMENTALITY

tible in significant ways with the rationality of biopower: the "health" of the markets implies the health of the population. In the final part I discuss the particular form of subjectivity—the *homo economicus*—it produces.

Truth

It might seem less than plausible to assert the significance of Foucault's thought in analyzing our contemporary economic reality, given that he is often read as a thinker who, to his detriment, largely ignored questions of political economy. Jeffrey Nealon, for example, has argued recently in his book *Foucault Beyond Foucault* that he has very little to say to today's readers about the economic present, which is not primarily geared towards a standardized and normalized mass society, but is instead supersaturated with neoliberal practices of individual self-creation.[4] According to Nealon, Foucault expanded most of his political and theoretical energy smoking out the hidden indignities of a form of governmental power that has decisively lost hegemony in the decades since his death, namely the welfare state. Making his thought relevant today would therefore require constructing a productive dialogue with contemporary Marxism. This, in turn, would mean acknowledging his affinities with certain of its core tenets—such as the persistence of class struggle—rather than viewing his relationship with Marxism as a wholesale rejection.[5]

It is my contention that while Marxist theory remains a pertinent analysis of many of the essential mechanisms of capitalism, it fails to identify what is specifically at stake in the rise of neoliberalism. I claim that Foucault's approach to neoliberalism is not only incompatible with Marxist analysis in crucial ways, moreover Foucault provides an original perspective precisely because he refuses to theorize it in terms of ideology and class struggle.

A traditional Marxist response would explain the hegemony of neoliberalism in terms of class antagonism. David Harvey, for example, argues in his influential analysis that the neoliberal turn was a deliberate and highly successful attempt to restore the power and the wealth of the upper classes.[6]

---

[4] Jeffrey Nealon, *Foucault Beyond Foucault: Power and Its Intensification since 1984* (Stanford: Stanford University Press, 2008), 11.
[5] Ibid, 81-82.
[6] Since the global neoliberal turn in the 1970s the income gap between the rich and the poor, or the ruling class and the working class, has considerably widened. Harvey shows that, whereas it narrowed considerably in most Western countries after the Second World War and stayed relatively stable for nearly three decades, since the neoliberal turn

Reagan and Thatcher placed themselves at the head of a class movement the determined aim of which was to restore its power. By capturing the ideals of individual freedom and turning them against the interventionist and regulatory practices of the state, capitalist class interests were able to protect and restore their position. It was the genius of neoliberal theory to provide a benevolent mask full of wonderful-sounding words such as freedom, liberty, choice and rights to hide the grim realities of this restoration of naked class power, locally as well as transnationally. The IMF and the World Bank functioned as conspiratorial centers for the propagation and enforcement of "free market fundamentalism" and "neoliberal orthodoxy"—forms of ideology with highly questionable scientific rigor.[7] For Harvey, resistance to neoliberalism thus requires unmasking the truth: we must expose it for what it truly is, namely a covert attempt to restore class privilege. We also have to rejuvenate class politics: class is not a meaningless or defunct category, but must remain the central conceptual weapon in the struggle against neoliberal hegemony.[8]

It is my contention that instead of treating neoliberalism as an ideological mask for a hidden truth we should respond to it on the level of the production of truth. Foucault was deeply suspicious of the notion of ideology. For him, the key philosophical question did not consist in drawing a line between what falls within the category of scientificity or truth and what comes under the suspicious label of ideology.[9] His interest rather lay in the production of truth in two distinct ways. He wanted to identify the political effects of truth and how they were produced historically. On the other hand, he also wanted to analyze the regimes of truth: the conditions that made it possible to utter true statements about governance or the economy, for example. Neoliberalism must be understood as a distinct regime of truth in this sense: its political ontology forms the conditions for making reasonable political judgments in today's world. Foucault's lectures chart this historical development, the genealogy that has

---

there has been an enormous spiraling of the levels of wealth in the top income categories. In the US, for example, the share of the national income taken by the top one percent of income earners fell from a pre-war high of 16 to less than eight percent by the end of the Second World War, and stayed close to that level for nearly three decades. The wealth that is now concentrated in the upper echelons of society has returned to a level that has not been seen since the 1920s. See Harvey, *A Brief History of Neoliberalism*, 15.
[7] Ibid, 21.
[8] Ibid, 202-203.
[9] See e.g. Foucault, "Truth and Power," in *Power/Knowledge: Selected Interviews and Other Writings 1972–1977 by Michel Foucault*, ed. Colin Gordon, trans. Colin Gordon, Leo Marshall, John Mepham, and Kate Soper (Brighton: The Harvester Press, 1980), 118.

established the neoliberal governmental regime of truth conditioning our current political practices.

Before explicitly turning to neoliberal governmentality, Foucault begins his lectures by tracing a line of investigation back to the eighteenth century. He shows how a new liberal form of governmental reason began to be formulated, reflected upon and outlined around the middle of the century, and how it found its theoretical expression and formulation in political economy.[10] Physiocrats such as François Quesnay in France had already given the economic domain a high degree of internal consistency, but it was essentially Adam Smith who established economics as a neutral, economic science. Through him the modern conception of the economy emerged as a separate sphere of society as well as an autonomous object of scientific knowledge in political history, and this was a highly significant development in terms of our conception of good government and, more generally, of our understanding of the political.

Foucault argues that with the development of political economy a new principle for limiting governmental rationality was established. While up to that point the law had functioned as an external limitation on excessive government, the new principle—political economy—was internal to the very rationality of government. This meant that government had to limit itself not because it violated the liberty or the basic rights of men, but in order to ensure its own success. Up until the middle of the eighteenth century there had been a multitude of imposed economic practices such as tax levies, customs charges and manufacturing regulations. All these were conceived as the exercise of sovereign or feudal rights, the maintenance of customs, or techniques for preventing urban revolt. With the birth of a new governmental rationality based on political economy the meaning of all these economic practices profoundly changed, however. From the middle of the eighteenth century it became possible to establish a reasoned, reflected coherence between them by means of intelligible mechanisms. This, in turn, made it possible to judge them as good or bad, not in terms of some legal or moral principle, but in terms of truth: propositions were subject to the division between the true and the false. According to Foucault, governmental activity thus entered into a new regime of truth (BB, 18).

---

[10] Foucault notes that the meaning of "political economy" (*économie politique*) oscillated between two semantic poles between 1750 and 1810–1820. Sometimes it aimed at a particular strict and limited analysis of the production and circulation of wealth, but in a broader and more practical sense, it also referred to any method of government that could produce the nation's prosperity (BB, 13).

The market had been a site of jurisdiction, both in the Middle Ages and in the sixteenth and seventeenth centuries, in the sense that it was invested with strict regulations ensuring that prices were fair, and that there was no fraud, theft or crime. It was also a site of distributive justice: the rules of the market ensured that the poorest could also buy things. Entry into a new regime of truth in the middle of the eighteenth century meant that the market no longer appeared, or had to be, a site of jurisdiction. It now appeared as something that obeyed and had to obey "natural," spontaneous mechanisms. The spontaneity was such that attempts to modify the mechanisms would only impair and distort them. The market thus became a site of truth—it allowed natural mechanisms to appear, and these permitted the formation of the right conditions for its proper functioning (BB, 30-31).

The market also essentially constituted the site of the veridiction of governmental practice: a good government now functioned according to truth rather than justice. This meant that limiting its reach also became increasingly a question not of rights, but of utility. Limiting the exercise of power by public authorities was no longer formulated in terms of the traditional problems of law or revolutionary questions concerning original rights and how the individual could assert them over and against any sovereign. From the beginning of the nineteenth century the key questions addressed to government were: Is it useful? For what purpose is it useful? Foucault claims that what fundamentally characterizes liberal government-ality is the idea that "governmental power is limited by evidence, not by the freedom of the individual" (BB, 62).[11]

The possibility of limitation and the question of truth are thus both importantly introduced into governmental reason through political economy. This is an extremely important moment in the history of govern-

---

[11] In addition to the two characteristics of the liberal art of government—the market as the site of truth and the limitation of governmentality by the calculus of utility—Foucault takes up a third feature: the globalization of the market as an objective. Until the middle of the eighteenth century economic activity was seen as competition over limited resources: there was only a certain amount of gold in the world, so as one state became enriched its wealth had to be deducted from the wealth of others. According to the new liberal art of government expressed by Adam Smith and the Physiocrats, competition under conditions of freedom could only mean that everybody profited. Competition in a free market would lead to maximum profit for the seller and, simul-taneously, minimum expense for the buyer. For the first time Europe appeared as an economic unit and the whole world gathered around it to exchange its own and Europe's products in the European market. This was not the start of colonization or imperialism, but heralded a new type of global calculation in European governmental practice: a new form of global rationality (BB, 56-57). A global market was thus set as an objective, even in this period.

mentality "since it establishes, in its most important features... a particular regime of truth which...is in fact still the same today" (BB, 18). Foucault's claim is not that at that moment in history politics and the art of government finally became rational, nor that an epistemological threshold had been reached on the basis of which the art of government could become scientific. He is rather arguing that governmental activity entered into a new regime of truth that conditioned what kind of claims could be reasonably made about it. This transformation was decisive for our current understanding of politics. It meant that all the questions formerly posed by the art of governing had to be reconfigured in order for us to be able to answer them in terms of truth or falsehood.

> At one time these amounted to the question: Am I governing in proper conformity to moral, natural, or divine laws? Then, in the sixteenth and seventeenth centuries, with *raison d'Etat*, it was: Am I governing with sufficient intensity, depth, and attention to detail so as to bring the state... to its maximum strength? And now the question will be: Am I governing at the border between...the maximum and minimum fixed for me by the nature of things...? (BB, 18-19)

Foucault's key claim is thus that our modern understanding of politics was constituted and limited by a particular liberal regime of truth, which established a new relationship between political power and economic knowledge. To sum up its essential features, it became possible, for the first time in history, to make scientific truth claims about economics and good governance. One of the most important ontological tenets of economic liberalism and neoliberalism is the doctrine of economic neutrality: economic facts are objective, universal and politically neutral. Political decisions have to be based on economic truths, which in themselves are understood to be politically neutral. [12]

---

[12] Teivo Teivainen (2002) calls "economism" the attempt to carry out state policies exclusively on the basis of economic analyses, which are understood to be neutral politically; see Teivo Teivainen, *Enter Economism, Exit Politics: Experts, Economic Policy and the Damage to Democracy* (London: Zed Books, 2002). He argues (17) that politically relevant decisions are increasingly made in institutions and contexts that are defined as economic and that are therefore outside of democratic decision-making. Democracy is restricted through the defining of various governance institutions and the issues they deal with as economic and using the doctrine of economic neutrality to produce a dichotomy between the economic and the political spheres. Examples include Central Bank independence, balanced budget amendments, exchange-rate rules as well as commitments to specific policy rules associated with trade and investment through international or regional institutions such as the International Monetary Fund (IMF)

This regime also implied that good government could not interfere with economic mechanisms. Because economic truths dictated that market mechanisms—Adam Smith's invisible hand—best ensured that the pursuit of private interests spontaneously led to the common good; it was therefore irrational to place such pursuits under political control. All possible market distortions had to be avoided to ensure the correct formation of prices, because only correct pricing effectively guided resource allocation towards efficiency, equity and stability. This meant that once something was defined as an economic question—such as the magnitude of the income gap between the rich and the poor—it was moved out of the political realm, which was understood as a realm that could cause needless interference in accordance with a set of political commitments and moral principles. Economic truths, on the other hand, could not be argued against politically without falling into irrationality.

This idea has reorganized our political ontology in carving out an autonomous realm of economy free of political interference. From a Foucauldian perspective the rise of neoliberalism must be understood as the culmination of a historical development that redrew the ontological boundary between economy and politics. Under neoliberal governmentality the autonomy of the economic sphere places strict limits on the realm of politics, such that economic knowledge must fundamentally guide and condition political power.

In terms of political resistance this means that the essential philosophical task is not to reveal the hidden truth about neoliberal economic theory and policy. More fundamentally, it is to ask, how has politics become a domain in which it is possible to make scientifically true claims about an increasing number of issues? Neoliberalism cannot be reduced to just another political belief that one is at liberty to adopt or discard. When it is understood as the extreme articulation of liberal governmentality it forms the current conditions for formulating political beliefs as such.

This means that the Left has not been duped by dubious ideological propaganda into accepting neoliberal economic policies: it has been defeated by truth. Truth poses a far more difficult political conundrum than ideology or

---

and the North American Free Trade Agreement (NAFTA). As a consequence, areas defined as economic or financial have been increasingly insulated from democratic parliamentary control. Although the insulation of some policy-making areas from democratic control and accountability is necessary and beneficial to the overall functioning of democracy, the danger is that the domain of democratic politics will become excessively narrow.

the restoration of class privilege because opposing it politically appears irrational. Margaret Thatcher, one of the most famous advocates of neo-liberalism, summed up the seeming inevitability of the neoliberal economic reforms in her slogan, "There is no alternative." The absurd question that the Left has had to face is: How can one resist economic truths politically?

Power

Some commentators have contended that the biopolitical societies, which began to take shape in the seventeenth century and crystallized in the extended welfare states of the 1960s and 1970s, have since collapsed: neo-liberal hegemony has brought the era of biopolitics to an end. Biopolitical care in the form of a tight control of populations has ceased to exist and globalization has proceeded largely without any biopolitical considerations for the health and happiness of individuals or populations.[13]

My aim in this section is to question such an interpretation and to argue that neoliberalism should be understood as a powerful mutation of bio-political governmentality. The fact that it has become the hegemonic model even in countries which traditionally had strong welfare states shows that its underlying values, at least in Europe, are not so much libertarian, but utilitarian. The neoliberal economic argument has won in the governmental game of truth organized according to the undisputed, biopolitical value of life: the aim of good governance is the maximal material wellbeing of the population. Only economic growth, a continuous increase in productivity, can deliver higher living standards for everybody and thus ensure the best care of life. My claim is that the rise of neoliberalism has meant that while the means for achieving this aim may have changed, the biopolitical end of maximal life has nonetheless remained the same.

Foucault's short discussion on biopolitics at the end of *The History of Sexuality, vol. 1* was followed in the subsequent year by the lecture series *The Birth of Biopolitics* (1978–1979). The very title suggests that the lectures were intended as an elaboration of the topic. However, their actual content appears to have nothing to do with biopolitics, and concerns economic

---

[13] See e.g. Mika Ojakangas, "Impossible Dialogue on Bio-power: Agamben and Fou-cault," *Foucault Studies*, No. 2 (2005): 5-28. Ojakangas argues that the fact that the era of biopolitics is coming to an end precisely at the moment when the nation-state is coming to an end suggests that the exercise of biopolitics presupposes sovereignty, if not *de jure* then at least *de facto*.

liberalism and neoliberalism instead. A quick look at the index reveals that the word biopolitics occurs in only four instances, and in two of these the context is an apology for the fact that Foucault had spent too long on other topics and had not been able to talk about it.[14] The lectures thus give no easy or conclusive answer to the question of how biopolitics and neoliberalism are related.

In the first lecture Foucault introduces biopolitics as the general topic of the series and gives a general characterization of its relationship to liberalism: the governmental regime of liberalisms must form the framework for understanding biopolitics. "It seems to me that it is only when we understand what is at stake in... this governmental regime called liberalism... will we be able to grasp what biopolitics is" (BB 21-22). In the course summary he again apologizes for the fact that the course ended up being devoted entirely to what should have been the introduction. He insists again, however, that biopolitical issues could not be understood as separate from the framework of political rationality within which they appeared and took on their intensity.

> This means "liberalism," since it was in relation to liberalism that they assumed the form of a challenge. How can the phenomena of "population," with its specific effects and problems, be taken into account in a system concerned about respect for legal subjects and individual free enterprise? In the name of what and according to what rules can it be managed? (BB, 317)

The demands of biopolitics thus posed a theoretical challenge to liberal governmentality, and biopolitics and liberalism formed a historical intersection: they were linked *de facto*, not *de jure*. Nevertheless, Foucault argues that liberalism fundamentally determined the specific form that biopolitics assumed in Western societies. Rather than being imposed by totalitarian systems of coercion, it has, for the most part, developed as a complex regime of power/knowledge in Western societies.

As I argued in the previous section, what characterizes liberal governmentality is the idea that there can be no sovereign in economics. Economic rationality is not only surrounded by, but also founded on the fundamental unknowability of the totality of the economic process: the invisible hand is invisible precisely because there can be no totalizing sovereign view. The

---

[14] See e.g. BB, 185.

sovereign has to respect the natural and inevitable mechanisms of the economy in order to ensure the maximal wellbeing of all.

Therefore, if we analyze politics at the level of political ontology, we find that the erosion of sovereign power, which is now often attributed to globalization, already began in the eighteenth century. The eighteenth century crucially saw the emergence of new economic experts whose task it was to tell the government the truth about the natural mechanisms that it had to manipulate or respect. The economists of the time were able to explain, for example, that the movement of the population to where wages were highest was a law of nature (BB, 16). The discovery of "natural laws" in the social sphere meant that the form that biopolitics assumed in modern societies was essentially tied to the power of experts—economic experts and others with privileged access to scientific truths about life.

It is my contention that neoliberal governmentality is thus not contrary to modern biopolitical governmentality; rather, their rationalities are deeply interwoven and compatible in the sense that they both rely on expert power. Liberal governmentality effected a shift to a regime of truth that emphasized the limitation of government according to truth, at the expense of a juridical framework, and paved the way for a modern biopolitical society of experts and managers of life at the expense of sovereignty.

Many of the biopolitical techniques and regulations that proliferated throughout the nineteenth century were implemented by the State. Biopolitics has historically developed in tandem with the modern nation-state, but it has also retained relative independence from it. It has developed and spread not only in welfare states, but also in substate and transnational institutions and contexts: welfare funds, private institutions and insurance companies, for example. The rapid reduction of the state in conjunction with the rise of neoliberalism has not led to the disappearance of biopolitical rationality. On the contrary, neoliberalism can be seen as its new hegemonic form. Neoliberalism has successfully advocated biopolitical values and ends: the right to one's body, to health, to happiness, to the satisfaction of needs. It has effectively eroded the domain that is considered internal to a sovereign community and thus has questioned the power of sovereignty as such. At the same time, it has correspondingly expanded the domain of the economic and this way extended and strengthened the rationality of biopower.

The methods and techniques of biopolitics have dramatically changed with the rise of neoliberalism, however. I will finish this section with a brief discussion of neoliberal social policy, which provides an illuminative example of this transformation of biopolitical techniques.

Foucault's historical account of neoliberalism is suddenly interrupted in lecture eight in which he takes up the contemporary political issue of social policy in France at the end of the 1970s. In my view this lecture is highly revealing of the stakes involved in his enquiry into liberal and neoliberal forms of governmentality. The transition to the neoliberal model was literally happening in front of his eyes as he was delivering his lectures.[15]

Foucault argues that France had adopted full employment rather than price stability, and the provision of social services rather than the balance of payments as its primary and absolute economic objectives after the Second World War. The reasons for the liquidation of these previous forms of economic priority towards the end of the 1970s were connected to the serious economic crisis that had hit the country at the beginning of the decade, and attributed by economic experts to insufficiently rationalized economic decisions (BB, 195).

This neoliberal turn in France had a dramatic effect on social policy. The arguments that the neoliberal economists advanced at the time have become all too familiar to us in recent decades: due to extensive social security labor is more expensive and work moves to countries such as China where labor power is cheap. International competition is distorted to the detriment of countries with the most extensive social insurance cover. This is again a source of rising unemployment. All are worse off. As Foucault formulates this central neoliberal doctrine: "There is only one true and fundamental social policy: economic growth" (BB, 144). Social justice can never be the aim of successful economic policy.[16]

Foucault notes how the German ordo-liberals' conception of social policy from the 1930s is reiterated almost word for word in the French social policy reform of the 1970s. There should be two systems that, as far as possible, are impermeable to each other: the economy must have its own

---

[15] Foucault's point was not that neoliberalism had been implemented wholesale. He noted that its diffusion in France had taken place on the basis of a strongly state-centered, interventionist and administrative governmentality in the context of an acute economic crisis. This meant that it involved a whole range of specific features and difficulties (BB, 192). Nevertheless, he claimed that the basic principles of neoliberal governmentality were clearly visible in the policies, speeches and writings of President Giscard d'Estaing and his political and economic advisers: the construction of an advanced social project had to go hand in hand with the construction of an efficient market economy.

[16] Foucault cites a report that appeared in 1976 in the *Revue française des affaires sociales*, written as a study-appraisal of thirty years of Social Security.

rules and the social must have its specific objectives. However, they must be decoupled so that the economic process is not disrupted or damaged by social mechanisms, and so that moreover the social mechanism has a limitation, a purity as it were, such that it never intervenes in the economic process as a disruption (BB 200).

Economy is a game and the essential role of the state is to set the rules and to ensure that they are duly followed, but it must never interfere with the game itself. The rules must be such that the economic game is as active as possible and consequently to the advantage of the greatest number of people. There must be only one supplementary and unconditional rule: it must be impossible for any of the players to lose everything and thus be unable to continue playing. This is a safety clause for the player, a limiting rule that changes nothing in the course of the game itself, but which prevents someone from ever dropping totally and definitely out of it (BB, 201).

Such a system is understood as the only guarantee that the economic mechanisms of the game—the mechanisms of competition and enterprise—will be allowed to function for the rest of society. A society formalized on the model of competitive enterprise will be able to exist above the threshold of absolute poverty: everybody will have to be an enterprise for themselves and their families. Below the threshold there will an assisted, floating and liminal population, which for an economy that has abandoned the objective of full employment will be a constant reserve of manpower that can be drawn on if need be, but which can also be returned to its assisted status if necessary (BB, 206). Hence, the only point of contact between the economic and the social is the rule of safeguarding players from being excluded from the game. Below a given level of income the state must pay an additional amount, even if it means giving up the idea that society as a whole owes services such as health and education to each of its members, and even if it also means reintroducing an imbalance between the poor and others, between those who are receiving aid and those who are not (BB, 203-204).[17]

What is at stake in neoliberal governmentality is thus not class antagonism—worker's rights and demands versus those of the capitalist class. It is based on a completely different social ontology: society is an economic

---

[17] Foucault notes that social benefits are thus not meant to modify the causes of poverty. They will never function on that level, only on the level of their effects. The contrast to a socialist policy is clear: a socialist policy is a policy of relative poverty, the aim being to alter the gap between the incomes of the wealthiest and the poorest. Relative poverty does not figure in any way in the objectives of neoliberal social policy. The only issue is absolute poverty, the threshold below which people are deemed not to have an adequate income to ensure that they have sufficient consumption (BB, 206).

game for self-interested individuals. Foucault argues that the principle behind the neoliberal understanding of the political community is an inverted social contract: all those who want the social contract and virtually or actually subscribe to it form part of society until such a time as they cut themselves off from it. In the neoliberal conception of society as an economic game there is no one who originally insisted on being part of it, and consequently it is up to society and the rules of the game imposed by the state to ensure that no one is excluded from it (BB, 202).

As I argued in the previous section, contesting neoliberal hegemony politically is difficult because it means contesting economic truths. As the example of neoliberal social policy shows, it appears to be equally difficult even if we attempt to move the debate into the realm of values. Contesting neoliberal hegemony has come to mean contesting the undisputed value of economic growth. The goal of good governance in modern biopolitical societies is the maximal material wellbeing of the population. Achieving this biopolitical objective in the neoliberal framework unfortunately implies the inevitable widening of the income gap. Questions of social justice have mutated into economic facts while the undisputed biopolitical ends have remained the same.[18]

The subject

Several commentators have noted how neoliberal governmentality can be viewed as a particular production of subjectivity: it produces an economic subject structured by different tendencies, preferences and motivations compared to the political or legal citizen of a disciplinary society or a society of sovereignty.[19] The political subject is understood as an atomic individual whose natural self-interest and tendency to compete must be fostered and enhanced. He or she is a fundamentally self-interested and rational being who will navigate the social realm by constantly making rational choices

---

[18] As William Connolly has argued, the principles of capitalist economy conflict with the principles of equality that underlie the welfare state; see Connolly, "The Dilemma of Legitimacy," in W. Connolly (ed.), *Legitimacy and the State* (Oxford: Blackwell, 1984), 227-31. The welfare state needs a growing economy to support its redistributive programs, but the structure of the economy is such that growth can only be achieved by policies that are inconsistent with the principles of justice that underlie those welfare programs.

[19] See e.g. Trent Hamann, "Neoliberalism, Governmentality, and Ethics," *Foucault Studies*, No. 6 (2009): 37-59, and Jason Read, "A Genealogy of Homo-Economicus: Neoliberalism and the Production of Subjectivity," *Foucault Studies*, No. 6 (2009): 25-36.

based on economic knowledge and the strict calculation of the necessary costs and desired benefits. The popularity of self-help guides and self-management manuals are seen as a symptom of this current, neoliberal understanding of the subject: individuals are solely responsible for a number of problems that were previously considered social or political issues.[20]

It must be pointed out that this neoliberal production of a new form of subjectivity is not a direct consequence of implicit or hidden ontological presuppositions concerning human beings, however. The metaphysical or anthropological question of whether human beings really are naturally self-interested and competitive is ultimately irrelevant. The crucial and fundamental presupposition is that in order for us to be able to provide a rational explanation for economic mechanisms we must treat them as if they are self-interested and competitive. The production of a new economic subject is a consequence of neoliberalism's political ontology: economic rationality must be the rationality of the entire society.

Foucault argues that the Chicago School took this goal to the extreme by eliminating the difference between the social and the economic. It was characterized by its use of market economy analyses to decipher relationships and phenomena that were previously thought to belong not to the economic but to the social or political realm. Economy was no longer one domain among others, with its own particular rationality, it was understood as the rationality of the entirety of human action.

The generalization of the economic form of the market to the whole of society functioned effectively as a grid of intelligibility and a principle of decipherment for social relationships and individual behavior. This schema made it possible to reveal in non-economic processes, relations and behavior a number of formal and intelligible relations. It became possible to generalize the economic form of the market throughout the social body, including relationships that were not conducted, and therefore not usually analyzed through monetary exchanges. An important example is the neoliberal analysis of human capital.

The theory of human capital developed by economists of the Chicago School such as Gary Becker and Theodore Schultz in the 1960s and early 1970s was an attempt to fill a gap in formal economic analysis by offering a unified explanation of a wide range of empirical phenomena that had either

---

[20] See e.g. Barbara Cruikshank, "Revolutions Within: Self-Government and Self-Esteem," in Andrew Barry, Thomas Osborne, and Nikolas Rose (eds.), *Foucault and Political Reason: Liberalism, Neoliberalism and Rationalities of Government* (Chicago: The University of Chicago Press, 1996).

been given *ad hoc* interpretations or had baffled investigators. Becker, for example, refers to well-known phenomena such as the fact that earnings typically increase with age at a decreasing rate, and that unemployment rates tend to be negatively related to the level of skill.[21] The idea of human capital explains such phenomena by treating behavioral choices such as education and on-the-job training as investments made in people. People enhance their capabilities as producers and consumers by investing in themselves. The many ways of doing this include activities such as schooling, training, medical care, vitamin consumption, acquiring information about the economic system, and migration.[22]

These investments result not just in some incalculable increase in the individual's wellbeing but also in a calculable increase in his or her income prospects. From the worker's point of view labor comprises capital: it is ability, skills that can be acquired at a cost. Human capital comprises both innate and acquired elements. While the innate elements are largely out of our control, the acquired elements are not. If we make educational investments we can become ability machines producing income (BB, 244).

The most striking example that Foucault discusses is the mother-child relationship (BB, 229-230, 243-244). A neoliberal economic analysis would treat the time the mother spends with the child, as well as the quality of the care she gives, as an investment that constitutes human capital and on which she can expect a return. Investment in the child's human capital will produce an income when the child grows up and earns a salary. Similarly, economic analyses of marriage could be read as attempts to decipher what is traditionally considered non-economic social behavior in economic terms. Social relationships could be considered forms of investment: there are capital costs, and returns on the capital invested.

The theory of human capital represents one striking example of the extension of economic analysis into a previously unexplored domain: it makes possible a strictly economic interpretation of a whole range of phenomena previously thought to be non-economic. Neoliberalism, understood as a specific form of governmentality, requires that economics can

---

[21] See Gary Becker, "Investment in Human Capital: A Theoretical Analysis," *The Journal of Political Economy*, Vol. 70, No. 2 (1962), Part 2: Investment in Human Beings, 10. See also Becker, *Human Capital: A Theoretical and Empirical Analysis with Special Reference to Education* (New York: National Bureau of Economic Research, 1964), and Theodore W Schultz, "Reflections on Investment in Man," *The Journal of Political Economy*, Vol. 70, No. 2 (1962), Part 2: Investment in Human Beings, 1-8.
[22] See e.g. Schultz "Reflections on Investment in Man," 2, and Becker "Investment in Human Capital," 9.

and must analyze human behavior and its internal rationality: theoretical analysis must bring to light the calculation through which an individual decides to allot his or her scant resources to this end rather than another. This means that the object of economic analysis ultimately becomes any conduct whatsoever that employs limited means for one end among others.

> And we reach a point at which the object of economic analysis should be identified with any purposeful conduct which involves a strategic choice of means, ways, and instruments: in short, the identification of the object of economic analysis with any rational conduct (BB, 268-69).

Neoliberalism thus forms a comprehensive schema for understanding social reality: all rational conduct, whatever it may be, ultimately comes under economic analysis. Economic interpretation of all human behavior is not only possible, it is understood to be the best way to make sense of it. Foucault quotes Gary Becker, who formulated this most strongly by noting that any conduct that responded systematically to modification in the variables of the environment, "which accepts reality," must be susceptible to economic analysis (BB, 269).[23] *Homo economicus* is anyone who accepts reality.

Foucault claimed that economics was therefore no longer primarily the analysis of economic processes and their historical or mechanical logic—which it had been for both Marx and Adam Smith, for example. It was the analysis of the strategic programming of an individual's activity (BB, 223). The worker was no longer present in the analysis only as an object—the object of supply and demand in the form of labor power—he or she was an economic subject. Whereas according to the classical liberal conception economic man, *homo economicus*, was a man of exchange, a partner in the process of exchange, in neoliberal terms he is an entrepreneur of himself. As Jason Read points out,[24] Marxists and neoliberals understand labor in fundamentally different ways: for Marx it was a sphere of exploitation while for the neoliberals the difference between labor and capital is effaced through the idea of human capital. Neoliberalism scrambles and exchanges

---

[23] Becker's groundbreaking work in economics demonstrated how a whole range of behavior was rational from an economic perspective, including phenomena such as altruism and addiction that were generally understood as exceptions to purely economic interests. When economic rationality was defined broadly enough individuals always maximized their welfare as they conceived it: altruism, for example, maximized utility when the welfare of others was the person's object of interest. See e.g. Gary Becker, *The Essence of Becker*, eds. Ramon Febrero and Pedro S. Schwartz (Stanford: Hoover Institution Press, 1995), 218-39, 329-42.

[24] Read, "A Genealogy of Homo-Economicus," 31.

the terms of the opposition between "worker" and "capitalist" by constructing a society in which everybody is a capitalist, an entrepreneur of himself. This means that any antagonism between classes can only ever be apparent because ultimately everybody wants the same thing: to succeed in their enterprise and to win in the economic game.

The rationality of neoliberalism thus potentially extends everywhere: the whole of society becomes a game in which self-interested, atomic individuals compete for maximal economic returns. The aim of neoliberal governmentality is to create social conditions that not only encourage and necessitate natural competitiveness and self-interest, but that produce them. As Foucault notes, the individual's life is lodged, not within the framework of a big enterprise such as the firm or the state, but within the framework of a multiplicity of diverse enterprises connected up to and entangled with each other. The individual's very life—his or her relationships with private property, family, household, retirement—must make him or her into a permanent and multi-faceted enterprise (BB, 241).

Neoliberalism reconfigures the line between public and private and between economy and society. It advocates competition as the dominant principle for guiding human behavior in society: competitiveness at all levels and at various scales of human activity—from the individual to the household, from the nation to the world economy—is paramount. It constructs a social order that safeguards competition in free markets in the knowledge that such an order is superior, not only economically but also morally and politically—the most conducive to securing its members' freedom and happiness. Individuals who do well in this competitive environment must accept this framework and act accordingly: make carefully calculated strategic choices between the most effective means, ways and instruments. They must make long-term and short-term investments in different aspects of their lives and acquire sufficient economic knowledge to be able to calculate costs, risks and possible returns on the capital invested.

Conclusion

To conclude, my aim was to show that neoliberalism is a much deeper and more complex phenomenon than a mere economic doctrine. It is a political ontology that fundamentally shapes our current experience of the world by forming its constitutive conditions. This entails a fundamental re-thinking of the tools of critical thought as well as of political resistance. Since neo-liberalism is not just another political program we cannot fight it solely with the traditional weapons of politics.[25] To put it simply, neoliberal govern-mentality reduces politics to a single question: according to the best avail-able economic analysis, what kind of political arrangement would ensure that the population is as wealthy as possible? The economic, but also bio-political terms in which this question is framed determine that it is difficult to resist neoliberal arguments with socialist demands for equality or workers' rights, for example. Foucault claimed provocatively that although liberal governmentality existed, socialist governmentality did not. Socialist politics had, therefore, to operate within the framework of liberal govern-mentality (BB, 92).

It is my contention that effective resistance requires advocating some version of radical politics that questions the very terms in which our political options are set. It requires attacks along all of the three axes of "truth," "power" and the "subject." We must question the political neutral-ity of economic knowledge and analyze the way in which economic truths produce political effects. We must also advocate the seemingly crazy argument that the maximal material wellbeing of the population is not necessarily the undisputed aim of good government. And finally, we must acknowledge that it is through us, our subjectivity, that neoliberal practices of governing are able to function.

Although some people claim that the financial crisis of 2008 brought neoliberal hegemony to an end, from a Foucauldian perspective it seems clear that such a feat would require a much more fundamental trans-formation of our political ontology than the rather superficial changes in economic policy that many Western countries have implemented recently. It would require a profound and radical revolution in our governmentality, in the way in which we understand politics and govern societies, and ultimately ourselves.

---

[25] Cf. ibid, 25.

Biopolitical Life

Catherine Mills

A decade ago, the term "biopolitics" referred only to a minor strain of Michel Foucault's genealogical work on prisons, sexuality and governmentality; today, it names a field of scholarship in its own right. However, the proliferation of scholarship on biopolitics has not resulted in greater conceptual clarity than was available from Foucault's limited discussions of the term. Indeed, with various competing theorisations of biopolitics in circulation—ranging from Foucault's own critical analysis of the liberal governance of the population, Agamben's rewriting of biopolitics as thanatopolitics, and recent fascinations with the possibility of an affirmative biopolitics—, the idea of biopolitics risks diversification to the point where it will have little critical force. This conceptual proliferation and consequent confusion has prompted some to reject the notion of biopolitics altogether.[1] In my view, though, the concept does important work in terms of identifying a particular rationality of politics, and therefore should be retained—though this should be on the condition of further conceptual clarification. The aim of this paper, then, is to try to make some headway toward the necessary clarification, especially in terms of the conceptions of life that are invoked in the debate.

The central claim of the literature on biopolitics is that modern politics is characterised by a tight connection between the operations of the state and the phenomena of life, such as health, death, reproduction and so on. Given this, it is surprising that the central problem of how life and politics become

[1] Paolo Virno, *A Grammar of the Multitude*, trans. Isabella Bertoletti, James Cascaito, and Andrea Casson (New York: Zone Books, 2004). Paul Patton, "Agamben and Foucault on Biopower and Biopolitics," in Steven DeCaroli and Matthew Calarco (eds.), *Giorgio Agamben: Sovereignty and Life* (Stanford: Stanford University Press, 2007).

combined to form a "biopolitics" has received relatively little attention. Even where it has received attention, the focus has primarily been on the nature of modern politics, a politics, it is claimed, that takes hold of and controls the phenomena of life. But this is a one-sided way of addressing the problem. For one may also ask, what is it about "life" that attracts the interests of governance and what, in life, allows it to be an object of political technique? In what ways do the phenomena of life provoke a biopolitics? Given the rhetorical centrality of the "bio" in biopolitics, it is striking that there is a widespread reluctance in the literature to approach the problem of life in more than a gestural or fantastic way. While the equivocations of the concept of life have undoubtedly been productive, the referent of the "bio" in the term "biopolitics" remains almost completely undisclosed. Instead, it is the dark background upon which the machinations of modern politics play out. In a sense, the ghost of Aristotle returns in virtually every attempt to theorise the relationship between life and politics.

It may seem strange to claim that the problem of life has fallen to the wayside in discussions of biopolitics, for is this not exactly what is most at stake in this literature? One could point to the proliferation of notions of life—from *nuda vita* or bare life,[2] to creaturely life,[3] and surplus life,[4] to name but a few. These exploit the manifold senses of the term "life," insofar as it is used to name a range of phenomena from mere animation to more specific areas of experience such as "public life" or "personal life" and so on. But to give one example of the obfuscation of "life," let me point briefly to Nikolas Rose's influential and widely read text, *The Politics of Life Itself.*[5] At no point does Rose give any account of "life itself," preferring instead to "explore the philosophy of life that is embodied in the ways of thinking and acting espoused by the participants in [the] politics of life itself."[6] Thus, in a manner that is characteristic of much biopolitics literature, what is under investigation here is not *life*, but *politics*. While Rose's approach certainly has its advantages insofar as it brings out the discourses at work in contemporary consumer genetics and medicine, it also has its dangers. In

[2] Giorgio Agamben, *Homo Sacer: Sovereign Power and Bare Life*, trans. Daniel Heller-Roazen (Stanford: Stanford University Press, 1998).
[3] Eric Santner, *On Creaturely Life: Rilke, Benjamin, Sebald* (Chicago: University Of Chicago Press, 2006).
[4] Melinda Cooper, *Life as Surplus: Biotechnology and Capitalism in the Neoliberal Era* (Washington: University of Washington Press, 2008).
[5] Nikolas Rose, *The Politics of Life Itself: Biomedicine, Power and Subjectivity in the Twenty-First Century* (Princeton: Princeton University Press, 2007).
[6] Ibid, 49.

particular, the term "life" tends to operate as a signifier without referent, almost infinitely encompassing and divisible, with the consequence that "life itself" is whatever is said about it and the operations by which life is managed and directed are seen as almost inevitably efficacious. A further danger of this approach, then, is that it elides the ways in which the phenomena of life might exceed and escape the ways in which people think about them, as well as the practices that strive to contain and improve them.

In this paper, I want to address the lacuna in recent literature on bio-politics to take the prefix "bio," which makes biopolitics a specific political rationale and form of organisation, seriously. To do this, I discuss the ways of thinking about "life" that have emerged in biopolitics literature, and through that, trace some of the parameters of "life" as a problem for thinking about politics today.[7] In the first section of the paper, I discuss the contributions to a philosophy of life suggested by Giorgio Agamben in his work on biopolitics, especially the idea of an absolutely immanent "happy life." I suggest that Agamben's strong resistance to biological conceptions of life limits the appeal of his work, since this tends to foreclose analysis of the "bio" of biopolitics. Following this, I turn to Roberto Esposito's recent book, *Bios*, in which he urges attention to the work of Georges Canguilhem as a starting point for a positive biopolitics that sees the norm as an immanent impulse of life. I use Esposito's discussion as a springboard for reconsider-ing the role of norms in Foucault's own work on biopolitics—especially in light of his essay on Canguilhem, in which he emphasises the productive capacity for error internal to life. I conclude that it is in the relationship of error and norms that the connection between life and politics may be made apparent. The reciprocal production of social and vital norms in the human as living being, and their specific conjunction in concerns such as population health, eugenics and new genetics, precipitates a biological politics that then extends into other domains of living. This point of view suggests that biopower is less a matter of controlling life that it is a matter of managing error—or rather, it is the former by virtue of the latter. It also highlights the way in which the biopolitical state is fundamentally reactive in relation to life.

---

[7] This forms a kind of prolegomenon to an approach to biopolitics that engages more thoroughly with theories of biology—not as a simple substitute for a theorisation of biopolitics, but as a way of generating a theorisation that is more able to engage with the biological microstructures of human life and their potential—but I do not attempt to develop such an approach here. Such a project would be situated at the conjunction of theories of biopolitics and recent interest in the philosophy of life, as well as a genealogy of modern biology.

"Absolute Immanence": Agamben

Giorgio Agamben's work has done a great deal to focus attention on the notion of biopolitics, and has also contributed much to contemporary reflection on the concept of life. Agamben himself suggests a number of different formulations for thinking about life, most notably the category of "bare life," which he sees as the principal subject of biopolitics, and its opposite, the post-biopolitical, even post-subjective, notion of "happy life." This latter notion can be seen as Agamben's most positive contribution toward current philosophy of life, and for this reason, I focus on it here.

In the essay "Absolute Immanence," Agamben notes that both Michel Foucault and Gilles Deleuze turn toward a discussion of "life" in the last of their publications during their lifetimes—entitled "Life: Experience and Science"[8] and "Immanence: A Life..."[9] respectively. This coincidence, he suggests, bequeaths to future philosophy the concept of life as a central subject, inquiries into which must start from the conjunction of Foucault and Deleuze's essays. While Foucault's essay—which is on the philosophy of life developed by his mentor, Georges Canguilhem—aims at "a different way of approaching the notion of life," Deleuze seeks "a life that does not consist only in its confrontation with death and an immanence that does not once again produce transcendence."[10] Insofar as these essays provide a "corrective and a stumbling block" for each other, they clear the ground for a genealogy that will, according to Agamben, "demonstrate that 'life' is not a medical and scientific notion but a philosophical, political and theological concept."[11] Such an inquiry would reveal the archaism and irrelevance of the various qualifications of life: animal life and organic life, biological life and contemplative life etc., and give way to a new conception of life that

[8] Michel Foucault, "Life: Experience and Science," in *Aesthetics, Method and Epistemology: Essential Works of Foucault 1954-1984, Vol 2*, ed. James Faubion (London: Allen Lane, 1998). This essay was initially published in 1978 as the introduction of the English translation of Georges Canguilhem, *Le normal et le pathologique* (Paris: Presses Universitaires de France, 1966), reprinted in Canguilhem, *The Normal and the Pathological*, trans. Carolyn Fawcett (New York: Zone Books, 1991). Another version of it was published in *Revue de métaphysique et le morale*, appearing in 1985, shortly after Foucault's death.
[9] Gilles Deleuze, "L'immanence: Une Vie...," in *Philosophie*, 47, No. 1 (September 1995). Republished as "Immanence: A Life," in *Pure Immanence: Essays on a Life*, ed. John Rajchman (New York: Urzone, 2001).
[10] Giorgio Agamben, "Absolute Immanence," in *Potentialities: Collected Essays in Philosophy*, ed. Daniel Heller-Roazen (Stanford: Stanford University Press, 1999), 238.
[11] Ibid, 239.

recognises beatitude—blessedness or happiness—as the "movement of absolute immanence."[12] Arguably, it is toward such a conception of life that Agamben's own philosophy aims: in proposing a typology of modern philosophy in terms of the thinking of transcendence (Kant, Husserl, Levinas and Derrida via Heidegger) and immanence (Spinoza, Nietzsche, Deleuze and Foucault via Heidegger), Agamben evidently positions himself as the philosophical heir of Deleuze and Foucault.

This is confirmed in his interpretation of Deleuze's notion of an absolutely immanent non-individuated life, which is the focus of Agamben's interest in "Absolute Immanence." Deleuze develops this idea through reference to Charles Dickens' story, "Our Mutual Friend," in which Riderhood wavers on the point of living and dying and compels unprecedented fascination and sympathy in witnesses to his predicament. Deleuze uses this story to develop a conception of a non-subjective or "impersonal" life, which is composed of "virtualities, events, singularities."[13] and which may be manifest in but is not reducible to an individual. Commenting further on the Dickens story, Agamben emphasises the way that this "separable" life exists in the indeterminacy between states of being such as life and death, which he describes as a "happy netherworld" that is neither in this world nor in the next, but between the two.[14] He goes on to cast the Deleuzian notion of a life of absolute immanence within the conceptual framework of biopolitics proposed in *Homo Sacer*, suggesting that impersonal life risks coinciding with the "bare biological life" of biopolitics. In Agamben's interpretation, Deleuze escapes this apparent declension by virtue of two related factors: first, the insistence on the "absolute immanence" of impersonal life, such that "*a life... ...is pure potentiality that preserves without acting*,"[15] and second, the connection between potentiality and beatitude, whereby the former is immediately blessed in lacking nothing. This means, "*[b]eatitudo* is the movement of absolute immanence."[16] The value, then, of reading Foucault and Deleuze's essays together is that it complicates both, such that "the element that marks subjection to biopower" must be discerned "in the very paradigm of possible beatitude."[17]

---

[12] Ibid, 238.
[13] Deleuze, "Immanence: A Life," 31.
[14] Agamben, "Absolute Immanence," 229. For further discussion of Agamben's interpretation of Deleuze's essay, see Catherine Mills, *The Philosophy of Agamben* (Stocksfield: Acumen, 2008).
[15] Agamben, "Absolute Immanence," 234.
[16] Ibid, 238.
[17] Ibid.

This conjunction and complication of Foucault and Deleuze contributes to the central role that the notion of happy life plays in Agamben's political philosophy, insofar as it points to a form of life beyond the biopolitical terrain of *bios* and *zoe*, of bare life and violence. Agamben makes clear his belief in the political necessity of such a conception of life in *Means Without End*, where he writes:

> The "happy life" on which political philosophy should be founded thus cannot be either the naked life that sovereignty posits as a presupposition so as to turn it into its own subject or the impenetrable extraneity of science and of modern biopolitics that everybody tries in vain to sacralize. This "happy life" should be rather, an absolutely profane "sufficient life" that has reached the perfection of its own power and its own communicability – a life over which sovereignty and right no longer have any hold.[18]

In this formulation, Agamben augurs a politico-philosophical redefinition of a life that is no longer founded upon the biopolitical separation of natural life and political life, and which is in fact beyond every form of relation insofar as it is life lived in absolute immanence. Happy life offers a kind of redemptive hope wherein the coming politics it grounds redeems humanity in the face of biopolitical annihilation. While a full exploration of the political and philosophical importance of the notion of happy life— including the plausibility of Agamben's intermingling of Foucault and Deleuze—is beyond the scope of this paper, two brief critical points can be made at this point.

First, Agamben's emphasis on the indeterminacy between life and death is underpinned by his formulation of potentiality and preserving without acting that is central to the "politics of pure means" that he advocates. This points to the theoretical rift between Agamben and other Italian political theorists such as Antonio Negri, who argues against the separation of potentiality and acting that Agamben emphasises. According to Negri and his long time collaborator, Michael Hardt, Agamben's mistake is to construe bare or naked life as fundamentally passive in relation to sovereign violence, singularly exposed without recourse or response. For them, Agamben's understanding of naked or bare life merely exposes "behind the political abysses that modern totalitarianism has constructed the (more or less)

---

[18] Giorgio Agamben, *Means without End: Notes on Politics*, trans. Cesare Casarino and Vincenzo Binetti (Minneapolis: University of Minnesota Press, 2000), 114-15.

heroic conditions of passivity" separated from action.[19] In contrast, Negri and Hardt claim in *Empire* that Nazism and fascism do not reveal the essential passivity of bare life so much as amount to an attempt to destroy "the enormous power that naked life could become."[20] There is significant debate around the political potential of a radical passivity, often characterised by the passive resistance of Herman Melville's Bartleby. Without entering into this debate here, Hardt and Negri's critique does highlight the impossibility within Agamben's conceptual framework of attributing an active force to life within a biopolitical context—redemption is only possible through the politico-philosophical conceptualisation of a life *beyond* biopolitics.[21]

Second, given the centrality of the relationship between life and knowledge in the essay by Foucault that Agamben mentions, we might well ask after his own relationship to biological knowledge. Or, in other words, provocative as his affirmation of a life of beatitude is, to what extent does it help to articulate the "bio" of modern biopolitics? Perhaps it is churlish or misguided to demand this from him, especially since it is precisely at a kind of deconstruction of the distinctions between nature and culture, or biology and politics, that his thought aims. Nevertheless, one can surely ask how successful such a deconstruction can be when one side of the opposition is left almost entirely obscure. The problem is that Agamben's sweeping claim that life is neither a biological nor medical concept forestalls engagement with the specificity of the ways in which biology and medicine have and do contribute to contemporary understandings of life. At this juncture, one might cite Georges Canguilhem, when he writes that:

> It is quite difficult for the philosopher to try his [or her] hand at biological philosophy without running the risk of compromising the biologists he uses or cites. A biology utilized by a philosopher—is this not already a philosophical biology, and therefore a fanciful one? Yet would it nevertheless be possible, without rendering biology suspect, to ask of it

[19] Michael Hardt and Antonio Negri, *Empire* (Cambridge, Mass.: Harvard University Press, 2000), 366; Michael Hardt, and Thomas L Dumm, "Sovereignty, Multitudes, Absolute Democracy: A Discussion between Michael Hardt and Thomas Dumm About Hardt and Negri's Empire," *Theory & Event*, 4, No. 3 (2000), §16.

[20] Hardt and Negri, *Empire*, 366-67.

[21] For a more extensive discussion of Agamben's political theory, including his debt to the work of Walter Benjamin as well as Hardt and Negri's critique, see Mills, *The Philosophy of Agamben*.

an occasion, if not permission, to rethink or rectify fundamental philosophical concepts, such as that of life?[22]

Of course, then, mere reference to a biologist does not in itself prevent fanciful philosophical biology: behind the complex discussions in Agamben's later book *The Open*, for instance, lies the radical theoretical biology of Jakob von Uexküll, whose account of the being in the world of animals provides a touchstone for Heidegger in *Fundamental Concepts of Metaphysics*.[23] Nor should the biological sciences simply be used to shore up a certain discursive authority that cannot be achieved solely through the humanities. The objectivist conception of life that focuses on genes and production of proteins and so on that dominates in the biological sciences today may well have a certain authoritative hold on contemporary ontologies of life, but it is insufficient as a way of thinking through the complication of life and politics. Indeed, as Roberto Esposito and others indicate, there is good reason to think that this approach to life has contributed to the mobilisation of the biosciences in biopolitics.[24]

The social and the vital: Esposito

In fact, of the various contemporary theorists of biopolitics, Esposito's increasingly influential work probably comes closest to providing a biologically refined understanding of life. In *Bios*, he shows that the German Nazi regime relied upon the expertise of biomedicine to justify and carry out its murderous plans in the camps and institutions such as T-4. The Nazi operations, he argues, were effectively a "biocracy," in which the legitimacy of the biomedical sciences gave strength to the political powers, and in return, the regime provided the bodies required for biomedical experimentation. From this characterisation of the negative biopolitical core of

[22] Georges Canguilhem, "Aspects of Vitalism," in *Knowledge of Life*, eds. Paola Marrati and Todd Meyers (New York: Fordham University Press, 2008), 59.
[23] There is an interesting theoretical genealogy here, for Uexküll's understanding of the animal within its world (summed up in the concept of *Umwelt*) is also significant for the functional biology of Canguilhem, as well as the neuropsychologist Kurt Goldstein, whom Canguilhem often references. Uexküll's work has also been rehabilitated in recent and emerging theoretical biology on biosemiotics. See, for example, Jesper Hoffmeyer, *Biosemiotics: An Examination into the Signs of Life and the Life of Signs*, trans. Jesper Hoffmeyer and Donald Favareau (Scranton: University of Scranton Press, 2008).
[24] Roberto Esposito, *Bios: Biopolitics and Philosophy*, trans. Timothy Campbell (Minneapolis: University of Minnesota Press, 2008).

Nazism, Esposito seeks an affirmative biopolitics uncontaminated by the thanatopolitics that emerges in modern politics. He outlines three axes along which what he calls the "immunitary dispositif" of biopolitics must be overturned. These are: the *double enclosure of the body*, the *preemptive suppression of birth* and the *normativization of life*.[25] I want to focus on the third of these here, for I think it is here, at the point of its greatest strength, that Esposito's thesis also reveals its particular weakness.

In the closing pages of *Bios*, Esposito argues, contra Agamben, that the Nazi regime was characterised by an absolute normativisation of life, such that this regime did not derive its power from the subjective decision in the shadow of the suspension of law, but rather, in the derivation of a normative framework from the very "vital necessities of the German people." The relation between law and life at stake in this, he argues, entails a double presupposition whereby the juridical norm presupposes the facticity of life, and life presupposes "the caesura of the norm as its preventative definition." Thus, he concludes that Nazism created a "norm of life," not however, in the sense that it adapted its own norms to the demands of life, but in the sense that it "closed the entire extension of life within the borders of a norm that was destined to reverse it into its opposite," that is, into death.[26] The problem for Esposito at this point is to suggest a way forward to a genuine politics of life or an affirmative biopolitics that breaks this deadly knot in which life and norm are entwined and mutually presupposed. He argues that attempts to distinguish more clearly between life and norm, such as in transcendental normativism and juris-naturalism, are unsatisfactory responses, however, since neither the absolutisation of the norm nor the primacy of nature can be considered external to Nazism. Instead, then, Esposito looks for resources in philosophical traditions that have emphasised the radical immanence of life and norm, and which in that way undermine the double presupposition that ties them together in Nazism.

Of these resources, he suggests that the theorisation of vital norms developed by Canguilhem may be especially valuable, since it allows for the "maximum deconstruction of the immunitary paradigm and the opening to a different biopolitical lexicon."[27] To reach this conclusion, Esposito references the radical vitalisation of the norm that Canguilhem proposes in his essay on the concepts of the normal and the pathological in the history of medicine. Here, he argues that life is internally and necessarily norma-

[25] Ibid, 138-145.
[26] Ibid, 184.
[27] Ibid, 191.

tive, since even at the simplest level "living means preference and exclusion."[28] Living necessarily involves polarities of valuation, such that an organism cannot be understood as indifferent to the environment in which it finds itself. Esposito goes on to emphasise that this means that disease and health are both normative states in that both indicate new forms of life for the organism, and moreover, reveal the normal functioning of the body. Conditions of disease or biological abnormality are not simply deviations from a fixed prototype of the normal: they are instead normative forms of a qualitatively different order. Similarly, to be "normal" is not to coincide with a pre-established norm, but rather, to be able to harness and maintain one's own normative power: to be normal is to be able to create new norms. In view of this radicalised immanence of life and norm, Esposito writes that "If Nazism stripped away every form of life, nailing it to its nude material existence, Canguilhem reconsigns every life to its form, making of it something unique and unrepeatable."[29]

There is much in this account of the productive power of Canguilhem's thinking around vital norms that I agree with. His analysis of the immanence of norms in life may well help to undermine the separation and mutual presupposition of the facticity of life and normative transcendentalism. Moreover, this analysis rejects an objectivist approach to life and emphasises instead the vital potential in life, in terms of the capacity to generate norms. However, it is exactly this productive power of the immanent normativity of life that points the way toward identifying several shortcomings with Esposito's approach. For this normative capacity, the power to create norms that inheres in life, is itself conditioned by the environment or milieu in which an organism finds itself. Thus, one point, which Canguilhem is wholly committed to, that Esposito tends to skip over is that an organism by itself is never normal—er, what can be considered "normal" is wholly dependent on the relationship between the organism and its environment. Canguilhem writes:

> Taken separately, the living being and his environment are not normal: it is their relationship that makes them such. For any given form of life the environment is normal to the extent that it allows it fertility and a corresponding variety of forms such that, should changes in the environment occur, life will be able to find the solution to the problem of adaptation... in one of these forms. A living being is normal in any

[28] Canguilhem, *The Normal and the Pathological*, 136.
[29] Esposito, *Bios*, 189.

given environment insofar as it is the morphological and functional solution found by life as a response to the demands of the environment. Even if it is relatively rare, this living being is normal in terms of every other from which it diverges, because in terms of those other forms it is normative, that is, it devalues them before eliminating them.[30]

Thus, life is inherently normative, in the sense that it aims at the restoration of functional or "normal" relations between an individual organism and its environment. And as this suggests, health is a "normal" situation, insofar as it indicates that the organism is normatively attuned to its environment and is able to meet the demands of it. Conversely, pathology or disease is the incapacity to meet those demands; but "the pathological is not the absence of a biological norm: it is another norm but one which is, comparatively speaking, pushed aside by life."[31] Norms are not only internally specific to the organism but vary across the conditions of its existence, when a normal condition is either disrupted by physiological changes or by changes in the demands that an environment places upon an organism such that it can no longer meet those demands. This means that the normal is never attained once and for all, since norms themselves are always subject to revision and regeneration.

This correction to emphasise the relationship between the organism and its environment may seem like a relatively minor interpretive point; but I want to suggest that it actually has important implications, two of which I will mention here. The first point goes to the fact that the environment that human beings are located in is necessarily social, and as such, cross-cut with the force of social norms. As Canguilhem suggests, human norms are "determined as an organism's possibilities for action in a social situation rather than as an organism's functions envisaged as a mechanism coupled with the physical environment. The form and functions of the human body are the expression not only of conditions imposed upon life by the environment but also of socially adopted modes of living in the environment."[32] This locatedness means that the "normal" is always an effect of a complex co-mingling and expression of vital norms in the midst of socially defined ways of living. Human life is never simply biological; and nor, for that matter, is it ever simply social or political. That Esposito leaves aside the necessary embeddedness of an organism in its environment means that he also risks obfuscating the ways that social norms cut across the vital norms

---

[30] Canguilhem, *The Normal and the Pathological*, 144.
[31] Ibid.
[32] Ibid, 269.

of the living human being. In doing so, his analysis runs close to the arguments of liberal eugenicists and transhumanists, who valorise the possible plurality of bodily norms that technologies of enhancement are supposed to engender, without consideration of the ways in which those possibilities are delimited in advance by social norms that are lived in often less than conscious ways.[33] In this, then, Esposito risks a version of the liberal fantasy of escape from the founding conditions of human existence.

The second point to make derives from this, for while the existence of human beings is fundamentally conditioned by social norms, it cannot be assumed that vital and social norms are conceptually equivalent. Rather, what needs to be taken into account is the disjuncture between vital and social norms, and consequently, what requires explanation is the means by which they intermingle. In other words, vital and social norms may well be empirically inseparable, but they are nevertheless analytically distinguishable. In the postscript to *The Normal and the Pathological*, Canguilhem argues that while physiological norms are immanent to the organism, social norms have no equivalent immanence. In a living organism, norms are "presented without being represented, acting without deliberation or calculation," such that there is "no divergence, no delay between rule and regulation." In contrast, rules in a social organization must be "represented, learned, remembered, applied."[34] Further, while biological norms are geared toward a functional end, social norms are not— speaking of the "health" of a society is metaphoric in a way that speaking of the health of a living body is not. The point of this is that forms of social organisation cannot be understood as analogous to organisms; nor, then, can social norms be simply derived from organic norms.

The point that social organisations are not analogous to organisms is significant for Canguilhem, because it allows him to avoid both the kind of socio-biology that derives social norms from biological norms, as well as a functional psycho-sociological version of adaptation that casts deviation from a social norm as abnormality. In light of this insistence on the exteriority of social norms, we would do well to qualify Esposito's thesis on the "vitalisation of the norm." While it is true that Canguilhem's work points the way to a new philosophy of life that emphasises the productive power of the living in terms of the capacity to create norms, he also resists a

---

[33] For further discussion, see Catherine Mills, *Futures of Reproduction: Bioethics and Biopolitics* (Dordrecht: Springer, 2011), especially Chapter 2, "Normal Life: Liberal Eugenics, Value Pluralism and Normalisation."
[34] Ibid, 250.

complete vitalisation of the norm, insisting on a more differentiated approach to norms and normalisation. This is important because while the exteriority—perhaps even transcendence—of social norms is indicated by the capacity to question those norms, it also opens them to such questioning and, ultimately, to transformation. In this regard, Esposito has notably little to say about one particular aspect of the productive power of the living that Canguilhem emphasises in the final pages of *The Normal and the Pathological*. This is the notion that life is characterised by an internal errancy, or capacity for error. Interestingly, it is this capacity for error that Foucault emphasises, suggesting that "Canguilhem has proposed a philosophy of error, of the concept of the living, as a different way of approaching the notion of life."[35] It is to Foucault's treatment of the capacity for error, and its implications for biopolitical life, that I now wish to turn. I have so far argued that the main theories of biopolitics are ill-equipped to articulate the prefix "bio" that gives the concept any specificity. Consequently, they fail to explain the active role that life plays in the operations of biopolitics. In the following section, I will suggest that Foucault's work provides resources for remedying these problems.

Biopolitics and error: Foucault

In the final chapter of *History of Sexuality*, Foucault makes his infamous argument that during the eighteenth century, the fundamental principle of Western politics changed from a sovereign power to a new regime of biopower, in which biological life itself became the object and target of political power. Biopower incorporates both disciplinary techniques geared toward mastering the forces of the individual body and a biopolitics centred around the regulation and management of the life of a new political subject, the population.[36] This new regime of political power operates according to

---

[35] Foucault, "Life: Experience and Science," 477.
[36] Michel Foucault, *The History of Sexuality, Volume 1: An Introduction*, trans. Richard Hurley (London: Penguin, 1981), 135-45. See also Foucault, *"Society Must Be Defended": Lectures at the College De France, 1975–76*, eds. Mauro Bertani and Alessandro Fontana, trans. David Macey (New York: Picador, 2003), and Foucault, *Security, Territory, Population: Lectures at the College De France, 1977–78*, trans. Graham Burchell (New York: Palgrave MacMillan, 2007), in which he explicitly discusses the idea of biopolitics. As this suggests, on occasion Foucault makes a distinction between "biopolitics" and "biopower," wherein the former term refers to the constitution and incorporation of the population as a new subject of governance, and the latter is a broader term that encompasses both biopolitics and discipline. Even so, he does not rigorously maintain it. I

CATHERINE MILLS

the maxim of "fostering life or disallowing it," and signals for Foucault the threshold of our modernity. It entails new forms of government and social regulation, such that power no longer operates through a violence imposed upon subjects from above, but through a normalising regulation that regularises, administers and fosters the life of subjects. In this new regime of power, power incorporates itself into and takes hold of the body of the citizen through the "normalisation of life processes."[37] Foucault concludes,

> [f]or the first time in history, no doubt, biological existence was reflected in political existence; the fact of living was no longer an inaccessible substrate that only emerged from time to time, amid the randomness of death and its fatality; part of it passed into knowledge's field of control and power's sphere of intervention.[38]

The field of biopower, then, is marked out by "the body [of the individual] as one pole and the population as the other," in a continual circuit of mutual presupposition and reference.[39] This characterisation of the emergence of biopower is perhaps so well known that some of its peculiarities are obscured through familiarity. In the following discussion, I want to call attention to some aspects of Foucault's account of biopower that may open ways of thinking beyond the current lacunae and confusions of the literature on biopolitics.

One of the key features of Foucault's account of biopower is the central role he gives to normalisation as a form of social and political regulation, suggesting at one point that "[a] normalizing society is the historical outcome of a technology of power centered on life."[40] As a technique of biopower,[41] normalisation is irreducible to the institutions and force of the law, and arises from the socio-political authority of statistics.[42] Interestingly,

---

use biopower in the discussion of Foucault to specifically indicate a technology of power that incorporates both discipline and a biopolitics of population.
[37] Georges Canguilhem, "On *Histoire de la folie* as an Event," in Arnold I. Davidson (ed.), *Foucault and His Interlocutors* (Chicago: Chicago University Press, 1997), 32.
[38] Foucault, *History of Sexuality*, Vol. 1, 142.
[39] Foucault, *"Society Must Be Defended,"* 253.
[40] Foucault, *History of Sexuality*, Vol. 1, 144.
[41] Normalisation is involved in both discipline and a biopolitics of population, though norms are mobilised differently, with different purposes, in each case. See Michel Foucault, *Discipline and Punish: The Birth of the Prison*, trans. Alan Sheridan (London: Penguin, 1979), 177-83, and Foucault, *Security, Territory, Population*, 57-63.
[42] On the history of statistics, see Ian Hacking, *The Taming of Chance* (Cambridge: Cambridge University Press, 1990). For a compelling account of the importance of statistics

Foucault suggests that normalisation works in opposing ways in discipline and a biopolitics of population. In the former, infractions of the norm are produced as a consequence of the prior application of the norm, insofar as the phenomenal particularity of an individual is itself identified and calibrated through the application of a norm. Normalisation produces individuals as the necessary mode and counterpart of the operation of norms, that is, as a material artefact of power.[43] In a biopolitics of population, Foucault suggests that norms are mobilised in exactly the opposite way, insofar as "the normal comes first and the norm is deduced from it." The biopolitics of populations, and the apparatuses of security that Foucault identifies as crucial to it, involves "a plotting of the normal and the abnormal, of different curves of normality, and the operation of normalisation consists in establishing an interplay between these different distributions of normality and in acting to bring the most unfavourable into line with the most favourable."[44] Given these different accounts of the modes of operation of normalisation at work in biopower, detailed studies of the mobilisation of norms in regard to specific instances of the management of life processes today are required. How, for instance, are norms formulated and applied in relation to human reproduction, including contemporary uses of genetic and reproductive screening technologies such as pre-implantation genetic diagnosis and ultrasound?

A precept of such detailed studies should be that in the case of ourselves as living beings, social and vital norms are simultaneously inseparable and irreducible; they do not determine each other, but neither can one be determined in the absence of the other. This condition of living in two worlds at one and the same time points to the ambivalence in the concept of life, where its meaning is often determined in its accompanying qualification, as, for instance, biological or social. As I noted earlier, Agamben sees such qualifications as themselves part of the operation of biopolitics, and because of this he resists any engagement with biological conceptions of life. Aristotle provides his starting point and target for thinking about life, especially the distinction between nutritive life (*zoe*) and political life (*bios*). However, this reveals a certain anachronism on Agamben's part. In Foucault's view, biopower is intimately related to the appearance of the biological in the sphere of politics; but biology, as a "discipline" or regime of

---

for Foucault, see Mary Beth Mader, *Sleights of Reason: Norm, Bisexuality, Development* (Albany: State University of New York Press, 2011).

[43] See Foucault, *Discipline and Punish*, 184.

[44] Foucault, *Security, Territory, Population*, 63.

truth, is a historically specific phenomenon and its categories and concepts cannot simply be read back into Aristotle. Moreover, to a large extent, the contemporary "molecularization" of life has overtaken the organic biology of the nineteenth century. It is undoubtedly true that discourses and practices of molecular life, which often tie individual and population identities to genetics through an integration with capital, are aligned with biopolitical strategies. However, this should not in itself render all engagement with biology suspect, for it may also be that less hegemonic conceptions of life emergent within contemporary theoretical biology provide ways of thinking beyond "recombinant biopolitics."[45]

Finally, and not unrelated to this, in a little analysed moment in his discussions of biopower in *History of Sexuality*, Foucault offers the caveat that one should not imagine that life has been totally administered and controlled by governmental techniques; rather, he states, life constantly escapes or exceeds the techniques that govern and administer it.[46] But what is meant by the term "life" such that what it refers to is able to escape the political techniques that seek to control it? What enables this moment of escape and in what form is it realized? At this point Foucault's discussion of the idea of an inherent potential for error in life developed in the thought of Canguilhem becomes important. In his short essay on Canguilhem mentioned at the start of this paper, Foucault argues that at the centre of the problems which preoccupy Canguilhem resides "a chance occurrence... like a disturbance in the information system, something like a "mistake," in short, "error"; Foucault states, "life—and this is its radical feature—is that which is capable of error."[47] As such, the error that is borne within life as its necessary potentiality provides the radical contingency around which the history of life and the development of human beings is twined for Canguilhem, which enabled him to identify and draw out the relation of life and knowledge. Foucault writes,

> if one grants that the concept is the reply that life itself has given to that chance process, one must agree that error is the root of what produces human thought and its history. The opposition of the true and the false, the values that are attributed to the one and the other, the power effects

[45] Michael Dillon, and Julian Reid, "Global Liberal Governance: Biopolitics, Security and War," *Millennium Journal of International Studies*, 30, No. 1 (2001).

[46] Foucault, *History of Sexuality*, Vol. 1, 143.

[47] Foucault, "Life: Experience and Science," 476. While this conception of life is more that of Canguilhem than it is of Foucault, it is possible to see the identification of a potential for error within life as at least a point of inspiration for Foucault.

that different societies and different institutions link to that division – all this may be nothing but the most belated response to that possibility of error inherent in life.[48]

Thus, it is through the notion of error that life is placed in a relation of contiguity and contingency with truth and structures within which it is told. "Error," or the inherent capacity of life to "err" both establishes the relation of life to truth and undermines that relation by disentangling man from the structures of truth and power that respond to the potential for error. Hence, "with man, life has led to a living being that is never completely in the right place, that is destined to 'err' and to be 'wrong.'"[49] From this point of view, the biopolitical state appears as simply the modern response to the possibility of error.

If this is so, the potential for error in life directs us to an important point about the operation of biopower, specifically, that the biopolitical state is necessarily and systematically *reactive*. The errancy internal to life constantly provokes the biopolitical state, forcing it to respond to the contingencies of the living and the phenomena of life. Today, a biopolitical state cannot *not react* to the provocations of life, even if that reaction proves to be a matter of disallowing life. Nevertheless, the ruse of biopower is to make it appear *as if* the state controlled and mastered life. There is a cliché that it is not so much the masters who walk their dogs, as the dogs who walk their supposed masters. To elaborate, it is not simply that humans tame dogs as pets—but rather, that somehow, dogs have managed to tame humans to such an extent that the latter will spend thousands of dollars and hours in keeping their pets alive and well. Similarly, we might consider the ways that life has demanded that the state care for it, has demanded—often quite successfully—that the state foster it by providing the conditions for its flourishing in manifold ways.

This is not to say that the biopolitical state does not also involve itself in the production of death—it evidently does; but when it does, it does so for the sake of the living. Foucault suggests at points that within biopower, death itself is relegated to the margins of political power: it is no longer a manifestation of the power of the sovereign, but precisely indicates the limits of power, the moment when life slips from the grasp of governance. Within biopower, death, Foucault suggests, "is outside the power relation-

[48] Ibid.
[49] Ibid.

ship," such that "power literally ignores death."[50] Foucault's comments here may be regarded as a transposed repressive hypothesis; in suggesting that "death is power's limit, the moment that escapes it; death becomes the most secret aspect of existence, the most 'private,'"[51] Foucault echoes the theoretical fallacy that he diagnoses with regard to sexuality. This then suggests the response that rather than being the limit of power, death is the means by which biopower functions—that is, it is precisely by recalling the risk of death, its immanence in life, that biopower operates, since it is the ever-present threat of death that justifies and rationalises regulatory intervention in the life of populations and individuals. Therefore, rather than attempting to eliminate or privatise death, biopower presupposes it for its operation; death is not the limit of biopower but its precondition. Against Foucault, we might say that it is not so much that "a relative control over life averted some of the immanent risks of death,"[52] but that an increasing control over death averts the immanent risks of life and permits its administration. This suggests, then, that biopower cannot be considered in terms of oppositions of "care" *or* "violence," "death" *or* "life"; instead, it establishes a mutually reinforcing relation between them: death is a precondition of living; violence is a precondition of care, and vice versa.

Conclusion

I have argued throughout this paper that contemporary literature on biopolitics has largely neglected to examine the concept of life that it continually invokes. Despite the proliferation of notions of life, the "bio" that marks the specificity of biopolitics as a rationality and technology of power fades into being little more than the obscure object of political techniques. The problem, though, is that to the extent that this is true, theoreticians of biopolitics risk participating in and reinforcing the very operations that they seek to diagnose. That is to say, the obfuscation of life, the apparent failure to conceptualise life as such, risks seeing it as merely an epiphenomenon of the state and of governance. This completely fails to see the active power of life itself. It mistakenly casts the state as productive of life, and therefore neglects the fundamental reactivity of the biopolitical state in relation to life.

---

[50] Foucault, *"Society Must be Defended,"* 248.
[51] Foucault, *History of Sexuality*, Vol. 1, 138.
[52] Foucault, *History of Sexuality*, Vol. 1, 142.

Towards an Affirmative Biopolitics
On the Importance of Thinking the Relations
Between Life and Error Polemologically

Julian Reid

"Who knows how to live well if he does not first know a good deal about war and victory?"[1] However complex the genealogy of the claim that war is the constitutive capacity for life is found to be within the counter-strategic tradition of modern political and philosophical thought (and it is deeply so), it is arguably to Nietzsche that we owe most for that understanding.[2] War was, for Nietzsche, not simply a primeval condition from which life must remove itself in order to secure the means for its peaceful flourishing, nor that instrument of the state which must merely be better deployed against other states, in order to secure the conditions for peace and security among its society, nor, for that matter, merely a mechanism by which the state secures itself from the disorder of the life it seeks to govern, but, rather, that which is ontologically fundamental for that life, and which, in being so, is formative of the conditions by which we might otherwise learn how to "live well" in struggle with powers seeking to stifle life of its capacities for such a knowledge. War is a fundamental capacity of life and in being so, is

---

[1] Friedrich Nietzsche, *The Gay Science*, trans. Walter Kaufmann (New York: Random House, 1974), 255.
[2] I have explored the depth and complexity of the function of war in determining the political ontology of what I have named the "counter-strategic" tradition of political thought in a variety of previous texts. Within this tradition I include Foucault, Deleuze, Virilio, Baudrillard, Negri, and Clausewitz, as well as Nietzsche. See especially *The Biopolitics of the War on Terror: Life Struggles, Liberal Modernity and the Defence of Logistical Societies* (Manchester: Manchester University Press, 2006); "Re-appropriating Clausewitz: The Neglected Dimensions of Counter-Strategic Thought," in Beate Jahn (ed.), *Classical Theory and International Relations: Critical Investigations* (Cambridge: University of Cambridge Press, 2006); and *Immanent War, Immaterial Terror, Culture Machine*, Issue 7 (2005): Biopolitics (with Keith Farquhar).

one which, once removed, leaves life in a condition of loss unto itself, and where it can encounter only the conditions for its own decadence. Thus is it that biopolitical modernity—in so far as it has been framed as a project dedicated to freeing life from war—has developed in violation of the constitutive conditions for a life lived well.[3] Construed, albeit in complexly different ways, as a project for the bringing of the war which informs life of all that which determines its wellness to an end—by declaring it over on account of having established the truth of life's desire and fitness for peace—biopolitical modernity could proceed only on the basis of the most reactionary form of violence to life. Indeed such is the grand error of the account of life on which biopolitical modernity, in its many different negational forms, has been based. For what is life, ultimately, without that capacity for a destructive shedding of those elements to be found within it, which do not strengthen but only serve to threaten it, limiting it, snuffing it of all that is most instinctive of it?

The violence done unto life—a violence on which biopolitical modernity has thrived—remains worthy of further exploration and critique. But these days we invoke the problematic of biopolitics in terms that are increasingly differentiated. Indeed we speak not just of biopolitics in the pejorative sense, through which we refer to the error of a violence that has been done to life on account of an erroneous understanding of what life is. But instead, and in the context of diagnosing the nature of that condition, we seek, now, an affirmative idiom of thought and expression for a politics that will constitute a different account of the necessary conditions for a life that might be lived well. One which, in its aspirations to make life live well, is capable of shedding its relations of subjection to regimes that would remove it of its capacity for war. Every regime of power entails its own particular account of who it is permitted to kill and how. An affirmative biopolitics is no different. It must also entail an account of how and what it needs to kill in order to make life live well. And so is it that in diagnosing the grand error on which biopolitical modernity has proceeded, which is to say the error through which we came to think that life could be lived well in simplistic accordance with a knowledge as to the truth of its will to peace, and in belief in the possibility of a renewed politics for the affirmation of life, so might we seek a restitution of an understanding of war as that which is the constitutive capacity for a life lived well. And so is it too that we might still want to read and return to Nietzsche for inspiration.

---

[3] See Reid, *The Biopolitics of the War on Terror*.

But before we can return to Nietzsche and affirm war anew as a constitutive capacity for a new kind of affirmative biopolitics so must we be prepared to problematize the relations between war and biopolitics a little further. Because biopolitical modernity itself has availed itself of a long, and equally complex, sedimented tradition of thinking war as the necessary condition for life too. The liberal subject of biopolitical modernity was never simply a subject the life of which must escape the condition of war in order to live, nor a subject the emancipation of which depends on how it can instrumentalize war as a means to living well. But a subject, not incomparable to the one that animates the Nietzschean imaginary, the life of which is itself said to be fundamentally conditioned by war. A subject the life of which can only be improved should it be able to destroy that rogue element of life found to err within it. A subject which is in a certain sense at war with its own living processes because it is within those processes that the conditions for its own capacities for error are to be located and destroyed.[4] A subject the life of which is composed of and by relations of war between generative and degenerative forces and out of which the relative capacities of that subject for truth and error will be decided. A subject which cannot live without engaging in a war that is not just constitutive of its life but operative within its life, against forces which are equally of its life as well as paradoxically found to be against its life. Thus is it, in other words, that in restituting an understanding of war as that which constitutes life we cannot seek simply to affirm the biological determinisms of the liberal tradition. We cannot simply affirm an idea of life as that which in struggling to improve itself through the modification of its existing constitution and adaptation to changing conditions, does so by the violent destruction of those errors which, produced by its own experimental desire to modify its forms, are found not to contribute to the betterment of itself. Instead affirming war as that which constitutes life requires a very careful delimittation of the variable subjects of life, and a very deliberate commitment to wage war in defense of the life that errs from the life which seeks the destruction of the life that errs, on account of its having determined the dangers which the path of error poses to the truth telling practices of its life-loving subject.

The difficulties of simply appropriating the concept of war from biopolitical regimes of power, re-deploying it so it might be thought as con-

---

[4] See Michael Dillon and Julian Reid, *The Liberal Way of War* (London and New York: Routledge, 2009).

stitutive of a life lived in contest of biopolitical regimes—as opposed to what is merely done to life on behalf of biopolitical regimes—has, of course, already been well exposed by Foucault in his *Society Must Be Defended* lectures. There Foucault delivered what remains in many ways a devastating critique of the problems that pertain to such an approach.[5] If we want to understand the paradox of how war came to be legitimized as what is done to life on behalf of life, we have to start by examining how modern political discourses, on the problem of the ways to enfranchise life, emerged in direct correlation with equally modern political discourses that sought to wage war against established forms of political power.

War, as Foucault demonstrated there, has long since been conceptualized within historico-political discourses as that force for the making of new forms of life which, in invoking it, may lay claim to the defense and promotion of their own ways of living. The subject which inhabits these force relations—the life of which is said to grow in accordance with the generative force of war—is a subject which must fight in order to survive, waging war in pursuit of the truth that he or she is. The seeking of the truth of the polemological subject is a decentering task.

It is the fact of being on one side—the decentered position—that makes it possible to interpret the truth, to denounce the illusions and errors that are being used—by your adversaries—to make you believe we are living in a world in which order and peace have been restored. The more I decenter myself, the better I can see the truth; the more I accentuate the relationship of force, and the harder I fight, the more effectively I can deploy the truth ahead of me and use it to fight, survive and win.[6]

In order to understand how war became conceptualized as a generative force of life, one has to begin not with Nietzsche, nor with the modern state, but with those historico-political discourses of antipathy towards the state that Foucault examines in "*Society Must Be Defended.*" In other words it is not that one can simply confiscate war from those traditions, which lend biopolitical modernity its reactive aspects by furnishing states with the right to kill life on behalf of life, in order to make war an affirmative capacity of those lives threatened by biopolitically sanctioned war. This is because the reactive conjugation of life with war emerged only as an inversion of those

---

[5] See Michel Foucault, *"Society Must be Defended": Lectures at the College de France 1975–76*, trans. David Macey (London: Allen Lane, 2003), and Julian Reid, "Life Struggles: War, Disciplinary Power and Biopolitics in the Thought of Michel Foucault," *Social Text* 86 (Spring 2006).

[6] See Michel Foucault, *"Society Must be Defended,"* 53.

discourses in which the relation of war to life was understood as something avowedly affirmative.

Inevitably this argument, if we accept it, has significant implications for any desire to restitute from Nietzsche an affirmative biopolitics based on his understanding of war as a constitutive condition for life. Elsewhere I have underlined how important it is to understand this critique, which Foucault developed with regard to the connections between war and discourses on political subjectivity in the context of his own growing hostility surrounding the direction in which many of his philosophical contemporaries were tending, those who thought that the fundamental question of politics was Nietzschean, namely how to assume war as a condition of possibility for the constitution and generation of resistance to biopolitical regimes of power.[7] Against such polemological theorizations of political subjectivity, Foucault posed the problem of war in starkly different terms. The problem being not how can war be restituted in generation of the political subject but when was it that war first came to be conceived as the source of political subjectivity?[8] Given the importance of the Nietzschean legacy for such polemological theorizations it is impossible not to read these lectures as an attempt by Foucault to distance himself from the Nietzschean conception of relations between life and war, and which had so inspired him up until that point.[9] The Nietzschean idea, which had previously animated Foucault, finds expression in the following, namely that there is "always something in the social body, in classes, groups and individuals themselves which in some sense escapes relations of power, something which is by no means a more or less docile or reactive primal matter, but rather a centrifugal movement, an inverse energy, a discharge...a plebeian quality or aspect."[10] It is precisely this polemological materialism that we can read Foucault locating, problematizing and attempting to think beyond in those lectures. There he identifies a deep complicity between such a position with biopolitical

---

[7] See Reid, *The Biopolitics of the War on Terror.*
[8] See Reid, "Life Struggles."
[9] It is also important to read the critique of the function of modern discursive relations between politics and war in framing biopolitical accounts of subjectivity in the context of his critique of the function of the discourse of war in shaping Nietzsche's theory of knowledge, and its importance for Deleuze. See Foucault, "Truth and Juridical Forms," in Foucault, *Power: The Essential Works 3*, ed. James D Faubion (London: Allen Lane 2000).
[10] Foucault, "Powers and Strategies," in Foucault, *Power/Knowledge: Selected Interviews and other Writings*, ed. Colin Gordon (New York: Pantheon, 1980), 138.

accounts of subjectivity which, as the modern age progressed, had become the sources of life sanctioned forms of war and violence against life.

In posing the question, what could be said of the life of the political subject, once it is divested of its grounding in ontologies of war and peace? Foucault had little guidance to give. Of course that may very well have been the point and on such basis this would make Foucault's political philosophy of life ultimately more Nietzschean than even Nietzsche himself. For once the conception of war as the constitutive condition for life was exposed for the metaphysical prop that it was—a truth of life which functions like all truths said of life to be its *dispositif*, and running thereby counter to Nietzsche's belief in the necessity to refuse all attempts to give life its truth —, so it befell the Nietzschean in Foucault to refuse the injunction to say of life anything at all.

But Foucault could not resist pursuing another different answer to this problem. Inspired by his teacher Canguilhem, and in writing the intro-duction to the latter's *The Normal and the Pathological*, Foucault ventured another beguilingly different but, also equally problematic definition of life. "In the extreme" and "at its most basic level" life is "what is capable of error" he stated.[11] And with the human, he argues, life produces its greatest error. And in being its greatest error its greatest work. For what is the human other than a "living being dedicated to 'error' and destined, in the end, to 'error.'"[12] It is error not war, he ventured, that "is at the root of what makes human thought and its history."[13]

Taking in consideration the breadth of Foucault's works, this is only a very minor essay from which I have quoted. But still, I would like to ask what follows when we attempt to think error itself as the constitutive capacity of life, in contrast to war. What follows for humanity and its politics when the human is conceptualized as that peculiar being dedicated to and destined for error? Does the conceptualization of error as the font of the life of the human provide us with a means to think and practice a politics of life without recourse to having to do violence to humankind and other forms of life? And does it therefore provide for more secure con-ditions on which to produce an affirmative biopolitics? Is a life that must err from itself in order to be itself more politically emancipatory than a life that must enter into conflict with itself in order to establish its own truth? How

[11] Foucault, "Introduction," in Georges Canguilhem, *The Normal and the Pathological*, trans. Carolyn R. Fawcett (New York: Zone Books, 1991), 22.
[12] Ibid.
[13] Ibid.

does this alternative formulation of the constitutive conditions for life impact upon questions of biopolitical subjectivity and its violences? Is the subject of error relieved of the imperative to kill?

There are some preliminary ways in which we might usefully compare the function of error in constituting the life of the human subject with that of conflict and it is by no means clear that the logos of error is distinct in actuality from the logos of war in this context. As can equally be said of the subject's capacity for war, error is not that which delays the arrival and securing of its truth, but that which makes new distributions of truth possible for the subject.[14] Errancy is not that practice of the subject which is to be overcome in order to secure the truth of its life, but rather that which constitutes its capacity to make truths which, rather than ossifying into transcendental forms, actually live, by becoming that which they are radically not; error. But there are also significant differences that come into view once error displaces war as the constitutive capacity for life. These differences are important not just in discerning the difficulties to be had in the break which Foucault was then attempting to make from Nietzsche, but in recovering the importance of life's capacity for conflict against other Foucault-inspired attempts to evacuate the ontology of war and enshrine life's capacity for error as an alternative and fundamental basis for an affirmative biopolitics. Here I am thinking especially of the work of Roberto Esposito, who, more earnestly than anyone else, has already attempted to develop an affirmative biopolitics out of an understanding of error as the constitutive capacity of and for life.[15]

In Foucault's works the implications of grounding life in an understanding of error as constitutive capacity in contrast with war were hinted at but never properly drawn out. But it is clear that for Esposito the fundamental value of such grounding is to be derived from how it impacts on the question of the relation of the biopolitical subject to its capacity for violence. An affirmative biopolitics, following Esposito, must be based on an account of a life lived in continuous openness to its errors such that "no part of it can be destroyed in favor of another."[16] Here in Esposito's Bios we confront an attempt to develop the concept of error in a way that radically reconfigures the relation of the biopolitical subject to the violence against life, which the Nietzschean ontology of war legitimates. But Esposito goes

---

[14] Ibid.
[15] See Roberto Esposito, *Bios: Biopolitics and Philosophy*, trans. Timothy Campbell (Minneapolis and London: University of Minnesota Press, 2008).
[16] Ibid, 194.

much further than Foucault by arguing that Nietzsche's insistence on the necessity of grounding life in war results not simply from his relation to the historico-political discourses exposed in "*Society Must Be Defended*," but is a product of an internal contradiction within his philosophical logic, and that as such, it befalls theorists of biopolitics, thinking with and beyond Nietzsche, to correct it.[17] "The open question," as he asserts, is "how to reconstruct the internal logic that pushes Nietzschean biopolitics into the shelter of its thanatopolitical contrary."[18] How to rid the Nietzschean subject of its erroneous belief that overcoming the killing and oppression of life, undertaken in the name of the preservation of its species being, requires making an error of the species, and inverting the war which has otherwise been conducted to eliminate humanity's errors, so that one might kill on behalf, and in promotion, of the life erstwhile said to have erred. Thinking life beyond Nietzsche must mean developing an account of a life, which, in the errors it makes, does not simply make new distinctions between the elements within itself that, on the one hand, strengthen it and those, on the other, which weaken it, nor a life which chooses to align itself with forces which err from, rather than supplement, what is already said to be true of it, but one which disavows itself of the rationalities upon which distinctions between truth and error are made, with a view to thinking life in terms that lead us beyond the discriminatory game of setting life against life.

But immediately we hit upon a problem with Esposito's construction of Nietzsche's biophilosophical logic. For the concept of error was by no means missing from the process through which Nietzsche assembled that logic. Indeed it was fundamental to the logic upon which Nietzsche constructed his own account of the relations between life and its capacity for war. One cannot simply counterpose error as an alternative foundation for the ontology of life to that of Nietzsche's concept of war, when error and war were mutually implicated within Nietzsche's philosophy of life to begin with. In this regard, Foucault was quite wrong to suggest that in giving primacy to error as a constitutive capacity of life, he was distinguishing his own position from Nietzsche's.[19] Defining life in terms of a capacity for error does not by any means initiate a break from Nietzsche's own understanding of life. Consider, for example, aphorism 307 from *The Gay Science*:

[17] Ibid, 93-101.
[18] Ibid, 98-99.
[19] Foucault, "Introduction," 22.

Now something that you formerly loved as a truth strikes you as an error; you shed it and fancy that this represents a victory for your reason. But this error was as necessary for you then, when you were still a different person—you are always a different person—as are all your present "truths" being a skin, as it were, that concealed and covered a great deal that you were not yet permitted to see. What killed that truth for you was your new life and not your reason: you no longer need it, and now it collapses and unreason crawls out of it into the light like a worm. When we criticize something, this is no arbitrary and impersonal event: it is, at least very often, evidence of vital energies in us that are growing and shedding a skin. We negate and must negate because something in us wants to live and affirm—something that we perhaps do not know or see as yet.[20]

As far as Nietzsche was concerned the production of error is fundamental for the human subject. Errancy is constitutive of truths that are necessary for the subject in question. But necessary only in the sense that they conceal from us our life in ways akin to that by which skin functions to protect us from the exposure of our flesh. Truth and skin as that which conceal all that we are not yet capable of seeing in ourselves. Nietzsche considered the error of truth to be fundamental for the capacity of the subject to secure itself from itself, while being also that which it must destroy in order to live. But in this sense we can see that Foucault, and Esposito after him, have sought a kind of erroneous reversal of the formulation of the relation between life and error posed by Nietzsche. Because error, for Nietzsche, is not that which life at its most affirmative is capable of, but that which, while necessary for the subject to survive, is also what must be killed in order for it to live. The error is both necessary, as well as expendable for the subject— and indeed error is that which *must* be expended in order for its life to break through the torpor of the truths it no longer needs. Thus error is, following Nietzsche, a kind of second rather than a first order capacity. While error is constitutive of the subject's production of truth, such error arises only as a "failure of the intellect,"[21] and it is to life that he assigns the task of ridding the subject of its failures.

Fulfilling the requirements of Esposito's affirmative biopolitics would thus mean not correcting a failure in his philosophical logic, but doing violence to his very deliberate and explicated understanding of the relations between error, life and war. An understanding as to the necessity of a sub-

---

[20] Nietzsche, *The Gay Science*, 245-6.
[21] Ibid, 196.

ject which in order to live well must struggle to destroy that element within itself that errs. For Nietzsche the life which errs, and which therefore must be destroyed, was not that life which threatens the constitution of truth within the subject but that movement of life within the subject that constitutes the desire to secure its truths. Thus it was the relation between life and error that, following Nietzsche, had to be conceptualized in terms of conflict. Error is the enemy in a conflict against which life is called to struggle. For life to affirm itself it must kill that which conceals it. The skin which conceals the flesh. In contrast with the position developed by Esposito we cannot affirm the life of the subject by insisting that it remain in a state of continual openness to its errors. That cannot be the condition for an affirmative biopolitics. Error is that capacity of the subject against which life must struggle in order to affirm itself.

For many, Esposito's attempt to follow Foucault and affirm in absolute terms the capacity of life for error will seem an attractive argument through which to stake out an alternative way of theorizing the contingency of relations between life and violence. It will inspire those who think that they can combat the liberal way of war without providing an alternative way of rationalizing war; and which is why, on the same basis, it is my view that it cannot be considered a sufficient ground on which to found a politics of resistance to liberal biopolitics. It is not possible to constitute an affirmative biopolitics, which does not in some sense rely on the constitutive capacity of life for war. Firstly because we can only think about life as error in the terms that Esposito urges upon us so long as we remain within an ethical dimension of thought. Esposito may claim that his philosophy of life as error is not an ethics and that we need to understand "friendship with the enemy not in an ethical sense, nor in an anthropological sense, but in a radically ontological sense,"[22] and that there is nothing altruistic in his account of error as destiny and the highest capacity of the human construed as a living entity. Error, he may argue, ought not to be considered an obligation which, in our humanity, we are required to meet if we are to "live well," but something which life does in spite of how else we might think we understand the nature of our status as living entities—or, even, how we may want to live, or think it necessary to live in order to do justice to life. But the following questions remain. How can we practice such a politics? Who is the political subject of life understood in terms of capacity for error? How does that subject differ from the political subject of life understood in terms of capa-

[22] Esposito, *Bios*, 107.

city for conflict? What kind of political subject can possibly bear such an affirmation not just of its errors, but also of its enemies? These are fundamental questions that Esposito still needs to unpack if he is to succeed in producing an affirmative biopolitics outside of, and beyond, the Nietzschean grounding of life in war and conflict.

In actuality we can only practice such a politics so long as we think about the problematic of the relation between life and war, not just within an ethical register, but an ethical register that is occupied by a sovereign form of life. For this is a problematic articulated from precisely that centered position of philosophical repose which only a sovereign form of life can occupy: a life which in recognizing the contingency of its hostilities, is able to decide to indemnify, and make a friend of its enemy. Imagining a life which grants a capacity for a mode of friendship with an enemy that develops from within its own body, Esposito is dedicated to conceptualizing the judgment of what is to be done with the error as that operation which is most normal of what we might call the "body normal." A body which, in order to be true to itself, must grant the error the status of what is most normal about it, so that thereby the production of error becomes understood as an expression of its capacity for life, rather than as a lack or remainder. So, identifying something vital in life's positing of an erroneous otherness, even though it's an otherness that threatens the "body normal," rather than something that at most might be tolerated in its difference from the norm. Therefore, it is a slightly more ambitious ethics than that posited by Habermasian theories of normativity, but nevertheless still an ethics articulated from the perspective of a "body normal." Ultimately, what is most problematic is Esposito's conceptualization of error as a product of the "body normal" rather than as that, which struggles against the violence of the body normal to find expression. At no point does Esposito venture to think the error from the perspective of the life named as erroneous. But for the erroneous the body normal is not that power which, in its decision not to destroy, gives life, but that which must be destroyed in order that it may live. For the error to constitute itself in accordance with its own powers of enunciation as something without any relation of subjection to the body normal, it must kill that body which names it error. Otherwise it remains subject to the discourse on which its erstwhile definition as erroneous depended.

Esposito's approach works only so long as we believe in the possibility and desirability of transforming political struggles between body normals and their errors into ethical ones by construing the problem of errancy purely from the perspective of the body normal. A body which in its possession of the power to suspend the norm in opening itself to its errors, may or may not decide to engage in hostilities with the life which emerged in subjection to it and which can only ever struggle against it for life. This is why the metaphor of pregnancy is so powerful for Esposito's thesis. It is a long way from Gilles Deleuze's much more faithful interpretation of Nietzsche, as well as Deleuze's far more political understanding about how the becoming of a life emerges always in context of a struggle with a body suppressing it.[23] A life which in order to be must always struggle to become against the being that may or may not withhold the power to determine it as erroneous. Such a life cannot rely naively on the benevolence of the body normal towards it because the power relations that connect it are, as Deleuze was always attentive to, so radically unequal. The body normal is not dependent on the error that it may or may not decide to grant life to. It can choose to destroy it and go on living should it choose to, even if the life that it lives may well be in some ontological sense impoverished by the reduction in potentiality that results from the destruction. Or it can choose to make a friend of the error and quite possibly go on living with it in an enriched condition of co-existence. But the life named error does not possess the same liberty with which to exercise judgment as to whether it ought or ought not to destroy that power to which it is subject. It cannot live in a relation of co-existence with the body normal should the latter choose to destroy it, and it cannot be certain that the latter will not decide to do so. Thus its only effective choice is to struggle against it with a view to achieving some measure of autonomy from it. It has to shed its relation with that body in order to be able to secure the conditions for its own becoming. Thus an affirmative biopolitics, thought from the position of error, is unthinkable without taking into account the necessary function of conflict in determining the capacity of error to become not normal, but true to itself.

Contrary to both Esposito and Foucault, error has to be thought not as a capability of life, or that which at its most basic life is, or what the human as a living entity is destined to produce or live in dedication to. It is not of the power of a subject dedicated to "living well" either to produce the error or to correct its relation with error by declaring its capacity to make error what is most true for it and thus absorb the practice of errancy within its own account of what a normal body can do. In order to affirm itself as other than error, the life that errs must be capable of condemning the body normal

---

[23] See Gilles Deleuze, *Nietzsche and Philosophy*, trans. Hugh Tomlinson (London: Athlone Press, 1983).

which not only has the power to determine it as error, but which, in attempting to redress that enunciative practice of naming the error, seeks to normalize it. The question of how to live in dedication to error is from the perspective of the life that errs a false problem. The problem is how to affirm itself without relation to the body in such a way that would claim it as that which it is dedicated to making live. Its self-affirmation depends not on the renunciation of error, nor simply the ability to think the production of error as its highest power, but to condemn as erroneous, the body which named it, and which in naming it, now claims possession of it, as the error which can be normalized. The body normal seeks to make a truth out of its error by naming it that which is most normal of and for it. But, for the life which is named as errant, true freedom lies in the ability not to be renamed as normal but to reconstitute the question of where errancy lies. The reconstitution of the question of "who has erred?" in determining one's life as errant, a deviancy from the normal, and thus subject to the renormalizing power of the body normal. The exercise of this expressly political power, a power which can only be exercised from the position of the errant subject, entails a will to see disappear the false problem of how to live in dedication to error, of how to destroy the power which thinks life in terms of errancy. How to condemn the false problem on account of which its life was named as erroneous with a view to destroying the enunciator of that problem so that it might realize its difference in kind from the body which once named it and now seeks to absorb it.

Only by reconciling and reaffirming these relations between error and war can we talk of the possibility of an "affirmative biopolitics." One that does not attempt to overcome Nietzsche's conceptualization of war as constitutive capacity for life by simply affirming error in its place. Such a blithe dedication to error is not an affirmation of life at all, but only of the subject's failure for life. An affirmation of the body which conceals that which actually lives within it. The skin which conceals its flesh. The truths which live in expense of its life. What lives in the subject is not just that element of itself which reconciles with the errors it makes, or conceives the ability to make errors as that which is most normal for it, but that power to conceive life as that which in being named the error emerges in destruction of the norm, and that which in condemning the norm, legitimates its destruction, willfully shedding it, in full exposure of itself. An affirmative biopolitics requires the conjugation of each of these terms and their cor-relate practices: life, error, and war. Overcoming the errors of the ways in which life has been conceived and practiced in an era of biopolitical dedication to the defense and promotion of life requires not dedicating ourselves to thinking life as capacity for error, but to thinking war as that capacity without which the erroneous cannot survive and prosper. Thus does it continue to be of necessity that we think biopolitics both affirmatively and polemologically. And so is it that Nietzsche remains an inspiration.

Biopolitics of Scale: Architecture, Urbanism, the Welfare State and After

Łukasz Stanek

Michel Foucault's notes on how, since the late eighteenth century, modern urbanism has been entangled with the biopolitical regime of security opens up the space for a general theoretical framework to account for the instrumentality of architecture and urbanism within and after the European welfare state, and requires posing the question of the historical specificity of this instrumentality in the post-war period.[1] This question can be addressed by reading Foucault in the context of the "scale debate" that has taken root principally in the fields of geography, sociology, and the political sciences. This suggests the possibility of theorizing the biopolitical project as a project of scalar organization of society, and urbanism as a project of biopolitics of scale, by which is meant the production of scales as historically specific frameworks of the biopolitical regime.

In this sense, architecture and urbanism of the post-war period need to be addressed by focusing on their instrumentality in the rescaling of socio-political processes which facilitated the shift from the consolidation of the welfare state to the processes of its increasing deconstruction initiated in the 1970s. This requires conceptualizing scale as socially produced material frame of social activity,[2] or, in the words of Erik Swyngedouw, as "the arena

---

[1] Michel Foucault, *Security, Territory, Population: Lectures at the Collège de France, 1977–78*, trans. Graham Bruchell (Basingstoke: Palgrave Macmillan, 2007); Michel Foucault, *The Birth of Biopolitics: Lectures at the Collège de France, 1978–79*, trans. Graham Bruchell (Basingstoke: Palgrave Macmillan, 2008).
[2] Neil Smith, "Remaking Scale: Competition and Cooperation in Pre-National and Post-National Europe," in Neil Brenner, Bob Jessop, Martin Jones, Gordon Macleod (eds.), *State/Space: A Reader* (Malden, Mass.: Blackwell Publishing, 2003), 228.

and moment, both discursively and materially, where socio-spatial power relations are contested and compromises are negotiated and regulated."[3]

Such a concept of scale goes back to regulation theory, itself developed in response to the crisis of Fordism and the welfare state in the course of the 1970s. This crisis is to be regarded as a crisis of one particular scale, namely the scale of the nation-state, which served as the frame of reference for achieving temporary compromises between competition and cooperation among social forces. This involved the correspondence between the national economy as the primary object of economic management; the nation-state as the primary political player; and national citizenship as providing the main definition of political subjectivity.[4]

Architecture, state, and the national territory

The nation-state was also the operative framework for spatial planning, a claim going back to such works as Walter Christaller's *Central Places in Southern Germany* (1933), which defined the number, distance and size of cities supplying the population of a given territory with services and commodities [see image 1].[5] This approach announced the promise of the welfare state as securing equal conditions of daily life for an entire population: a promise to be carried out by the distributive functions of post-war architecture and urbanism, charged with the task of allocating housing, transport, education, culture, and leisure. In architecture discourse from the late 1940s, this resulted in a debate surrounding the "greatest number," and concerned architects on both sides of the Iron Curtain. Thus the identification of the "society of the average man" (*société de l'homme moyen*) with "the problem of the Greatest Number" (as developed by the French architect Georges Candilis[6]) could be juxtaposed to that of Oskar Hansen's socialist Poland. Hansen, like Candilis a member of the Team 10, argued that only in socialism can the "the problem of the great numbers" be resolved.[7]

---

[3] Erik Swyngedouw "Neither Global nor Local: 'Glocalization' and the Politics of Scale," in Kevin R. Cox (ed.), *Spaces of Globalization: Reasserting the Power of the Local* (New York: The Guilford Press, 1997), 140.

[4] Neil Brenner, Bob Jessop, Martin Jones, Gordon Macleod, "Introduction: State Space in Question," in *State/Space*, 4.

[5] Walter Christaller, *Central Places in Southern Germany* (Englewood Cliffs, New Jersey: Prentice-Hall, 1966).

[6] Georges Candilis, "L'esprit du plan de masse de l'habitat," in *L'Architecture d'aujourd'hui* 57 (December 1954): 1.

[7] Oskar Hansen, "Linearny System Ciągły," *Architektura* 4/5 (1970): 125.

Hansen's project of the Continuous Linear System, drawn during the 1960s and early 1970s, and consisting of four large settlement strips stretching throughout Poland—from the mountains to the seaside—can be seen as an iconic expression of architecture addressing the national territory as a whole [see image 2]. This subscribed to the official discourse of socialist Poland about the "return" of Upper Silesia and Western Pomerania to the "mother country" after the Second World War. The theme of territorial integrity was just one among multiple links forged between the socialist state and Hansen's project, for which the planned economy and centralized building industry were essential premises. This radical reformism apparent in Hansen's work makes the Continuous Linear System a pedagogical project, directed against "real existing modernism."[8] The project was based upon an empirical analysis of the sites, which were mapped according to a method conceived by Hansen, and resulted in several detailed designs, some of which reached the stage of execution drawings. The project aimed at optimizing circulation on the level of the state and at delineating specific scales within the country as a whole. The principal criterion for this delineation were the quotidian practices of the inhabitant, who was granted the "right" to an urban experience, with all its heterogeneity and intensity. This was particularly perspicacious in the example of the "Masovian strip," which consisted of a cluster of functional strips intersected by people on their daily route to work. Similarly, in the area of the Western strip—starting in Upper Silesia—everyday experience was to be defined by all overlapping scales of the project, starting with individual houses, constructed by self-organized cooperatives of inhabitants, and ending with the view on the broad landscape from the terraced structures conveying infrastructure, provided by the state [see images 3-4]. In Hansen's words, "the classless, egalitarian, non-hierarchical character of the housing form for the society in the Continuous Linear System [...] should make legible to everybody his dependence on the collective and the dependence of the collective on the single person."[9]

Hansen's project is inscribed upon two centuries of a continuous interchange between biopolitics, architecture and urbanism. In the words of

---

[8] Cf. Łukasz Stanek, "Miastoprojekt Goes Abroad: Transfer of Architectural Labor from Socialist Poland to Iraq (1958–1989)," *The Journal of Architecture*, Vol. 17, No. 3, 2012: 361-86; Aleksandra Kędziorek & Łukasz Stanek, "Architecture as a Pedagogical Object: What to Preserve of Przyczółek Grochowski Housing Estate by Oskar & Zofia Hansen in Warsaw?," *Architektúra & urbanizmus* (forthcoming).

[9] Hansen, "Linearny System Ciągły," 135. See also Oskar Hansen, *Ku formie otwartej*, ed. Jola Gola (Warsaw: Fundacja Galerii Foksal, 2005).

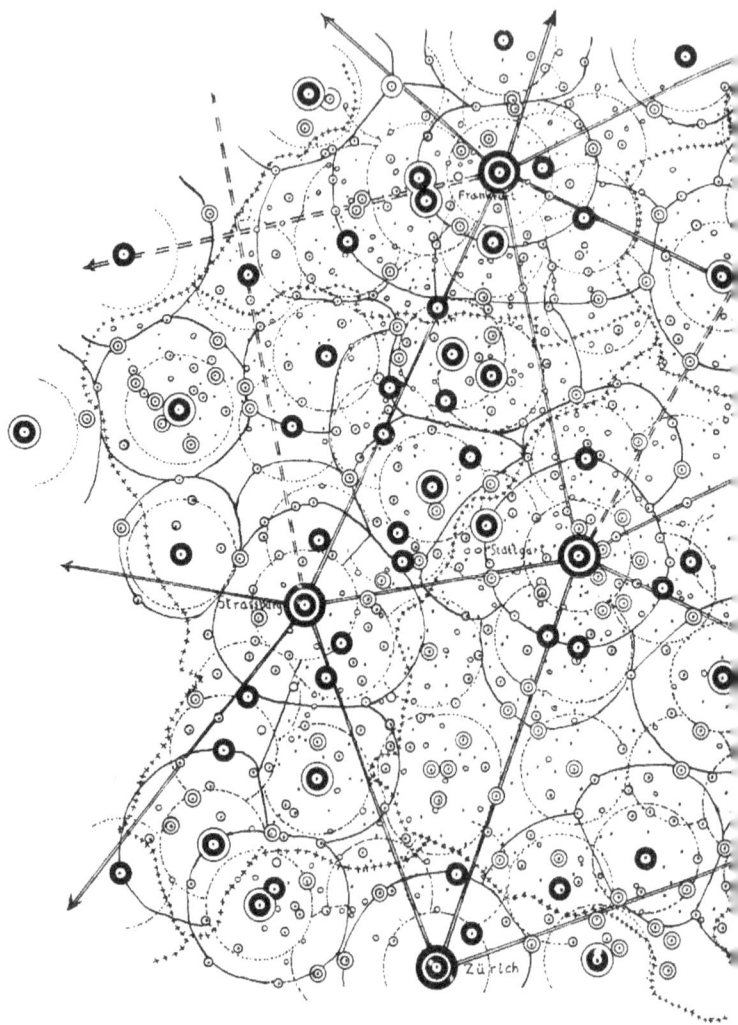

◉	L-Ort	⊘	K-Ort
◎	P-Ort	●	A-Ort
◉	G-Ort	·	M-Ort
◉	B-Ort		

Karte 4
Das System der zentralen Orte in Süddeutschland

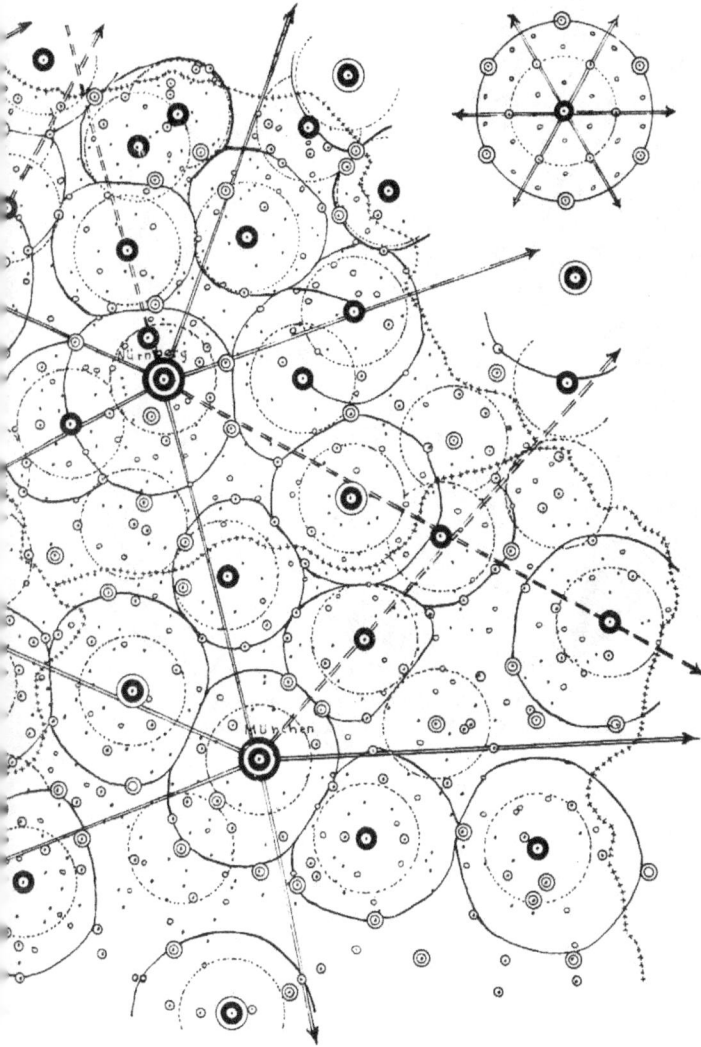

Rationales Schema der zentralen Orte ·

.... 21 km-K-Ring (schematisch)

———— Ring der B-Orte (normal 36 km)

++++++++ Grenzen der L-Systeme

===== L-Richtungen 1. Grades

== == L-Richtungen 2. Grades

Image 1: Walter Christaller, "The system of central places in Southern Germany" (1933), in: Walter Christaller, *Die zentralen Orte in Süddeutschland: eine ökonomisch-geographische Untersuchung über die Gesetzmässigkeit der Verbreitung und Entwicklung der Siedlungen mit städtischen Funktionen* (Darmstadt: Wissenschaftliche Buchgesellschaft, 1980).

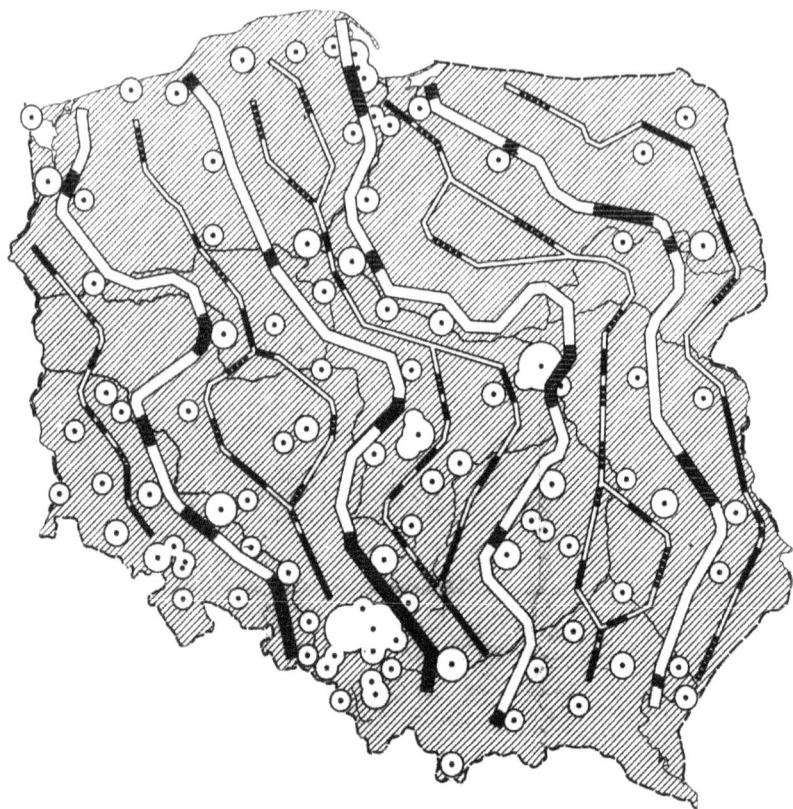

Image 2: Oskar Hansen, "Poland's Development Concept" (1977), in: Oskar Hansen, *Towards Open Form/ Ku formie otwartej* (Warszawa: Fundacja Galerli Foksal, Frankfurt am Main: Revolver, 2005)

Image 3: Oskar Hansen, "Western Belt (part II, 1976)", Multifunctional housing zone, postindustrial area (model) in: Oskar Hansen, *Towards Open Form/ Ku formie otwartej* (Warszawa: Fundacja Galerli Foksal, Frankfurt am Main: Revolver, 2005)

Image 4: Oskar Hansen, "Western Belt (part II, 1976)", Multifunctional housing zone (model) in: Oskar Hansen, *Towards Open Form/ Ku formie otwartej* (Warszawa: Fundacja Galerli Foksal, Frankfurt am Main: Revolver, 2005)

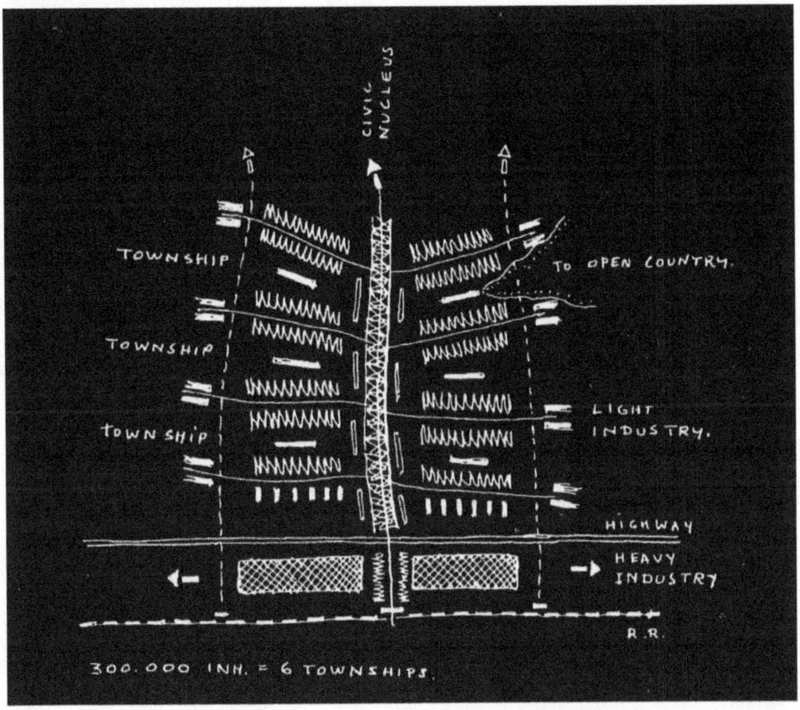
CIVIC NUCLEUS
TOWNSHIP
TO OPEN COUNTRY.
TOWNSHIP
LIGHT INDUSTRY.
TOWN SHIP
HIGHWAY
HEAVY INDUSTRY
R.R.
300.000 INH. = 6 TOWNSHIPS.

Image 5: José Luis Sert, "Settlement scheme", in: José Luis Sert, "The Human Scale in City Planning", in: Paul Zucker (ed.), *New Architecture and City Planning* (New York: Philosophical Library, 1944).

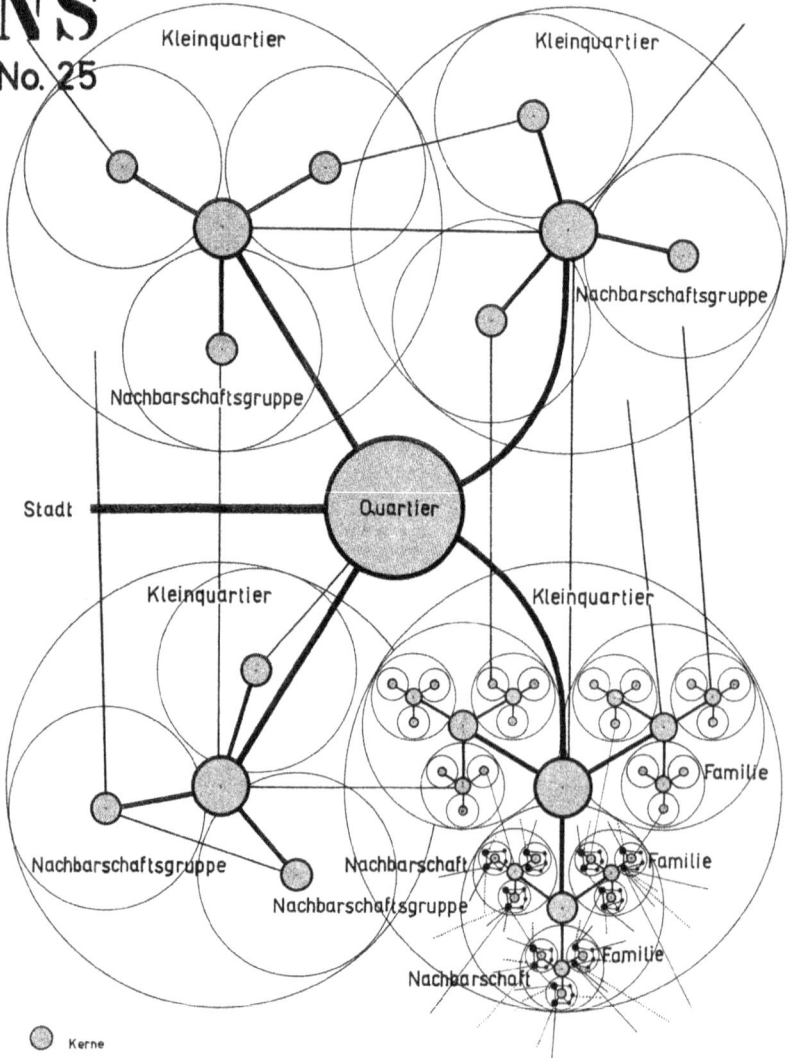

Image 6: "Diagram of human relationships in the city", in: *Fachgruppe Bauplanung der Studiengruppe 'Neue Stadt'*, headed by Ernst Egli, "Projekt einer Studienstadt im Raume Otelfingen im Furttal, Kt. Zuerich" (1958–1963).

Foucault, "from the eighteenth century on, every discussion of politics as the art of government of men necessarily includes a chapter or a series of chapters on urbanism, on collective facilities, on hygiene, and on private architecture."[10] As Sven-Olov Wallenstein has argued, if the introduction of such typologies as the hospital in the late eighteenth century can be seen as essential for the "emergence of modern architecture" it is precisely because in these structures "the idea of the building as an isolated object [was replaced] with a variable and flexible facility corresponding to the fluctuating needs of the population as a whole and entailing the introduction of 'public hygiene' as a new type of discursive object."[11] This logic is consistent with the institutionalization of urbanism by the eighteenth century as a biopolitical instrument, which takes as its proper scale of intervention the territorial circulation of people, commodities, money, orders, and crime, rather than the bounded space of a city. The functionality of urbanism in the management of a given population, as well as the distribution of risk according to an empirically accounted and statistically controlled norm, complements thereby Foucault's comments on the architecture of hospitals and prisons, revealing that the instrumentalization of architecture and urbanism within regimes of security went hand-in-hand with the development of disciplinary techniques. In this sense, Foucault's argument that modern biopolitics does not simply replace, but rather complements techniques of sovereignty and discipline, suggests that biopolitics is always already multiscalar, since it operates both as a production of the collective body of the population, as well as a production of individual disciplined bodies.

The multiscalar character of state agency came to the fore from the 1970s onwards, in the course of such interrelated processes as the increasing internationalization of economic relations; the resurgence of regional and local economies; the growing rejection of "overloaded" governments; the crisis of US hegemony in the international order and the increasing mobility of very large numbers of migrants across national borders.[12] These processes coincided with the introduction of new institutions, projects and struggles on multiple scales, relating to processes of globalization, but at the same time strengthened by the emergence of regional, local and urban

[10] Michel Foucault, "Space, Knowledge, and Power (interview with Paul Rabinow)," in Michael Hays (ed.), *Architecture Theory Since 1968* (Cambridge Mass., MIT Press, 1998), 430.
[11] Sven-Olov Wallenstein, *Biopolitics and the Emergence of Modern Architecture* (New York: Princeton Architectural Press, 2009), 33.
[12] Brenner et al, "Introduction," 1-26.

scales as increasingly important platforms for the "governance by community."[13] Specific to these changes is that social practices are operative not
only on many scales, but also across several scalar regimes; this is particularly the case with architecture, which can be understood today only as a
product of negotiation and struggle between global, regional and local
actors, for which European large-scale urban development projects are a
particularly well-researched example.[14]

In the course of these struggles across shifting hierarchies of scales, inter-
connected in broader, often-changing interscalar ensembles, the concept of
scale was redefined as a political concept: as products of economic, social
and political activities, scales became contested dimensions of social
practice. It is by conceptualizing, representing and organizing these dimensions that architecture and urbanism contribute to the politics of scale. The
focus on this contribution in the course of the twentieth century suggests a
tendential change from conceptualizing scales as discrete, self-contained
bounded spaces towards a topology of scalar systems in which the identity
of an element is defined only through its relations with other elements
within that system. In this sense, the transformation beyond the regime of
accumulation and the mode of regulation specific to the post-war European
welfare state was paralleled by a shift in conceptualizing urban scales from a
nested sequence of settlements (defined by a specific number of people,
functions, forms, and affective modalities of social bond) to an ecological
system of in-between spaces. This feeds into the argument of several
authors who have identified the realm of the in-between as the paradigmatic
site of biopolitics after Fordism, described by Antonio Negri as a site of
struggle between "the biopolitical exploitation of life" and a "resistance [...]
expressed in the experimental practice of an interstitial space."[15]

[13] Nikolas Rose, "Tod des Sozialen? Eine Neubestimmung der Grenzen des Regierens,"
in Ulrich Bröckling, Susanne Krasmann, and Thomas Lemke (eds.), *Gouvernementalität
der Gegenwart: Studien zur Ökonomisierung des Sozialen* (Frankfurt am Main: Suhrkamp, 2000), 72-109, See also Vol. 31 of *Dérive: Zeitschrift für Stadtforschung* (2008).
[14] Erik Swyngedouw, Frank Moulaert, and Arantxa Rodriguez, "Neoliberal Urbanization
in Europe: Large-scale Urban Development Projects and the New Urban Policy," in
Antipode 34 (2002): 547-82.
[15] Antonio Negri, Constantin Petcou, Doina Petrescu, and Anne Querrien, "What Makes
a Biopolitical Space?" in *Urban Act: A Handbook for Alternative Practice*, ed. Atelier
d'architecture autogérée (Paris: aaa/ PEPRAV, 2007), 290-306, citation on 292.

Bardet, Sert, and Egli: Scale as a discrete threshold

A series of cross-cultural moves cutting through the history of architecture and urbanism during the twentieth century—from Gaston Bardet, José Lluís Sert, Ernst Egli to the Team 10—would provide case studies for an investigation into this development in the conceptualization of scale. For Bardet, a major figure of French urbanism from the mid-twentieth century and a supporter of the catholic group "Economie et humanisme," urbanism is understood as the "science of human agglomerations," aiming at inter-relating social and spatial morphologies. In his paper "Community Scales in Urban Agglomerations" (1943), Bardet distinguishes six scalar community levels [*échelons communautaires*]: patriarchal, domestic, parochial, urban, metropolitan regional and metropolitan capital. For Bardet, the first three levels are unable to generate cultural values, and the last two are destroying entirely spiritual and traditional values. For these reasons the "urban" level of 5,000-15,000 families is optimal in Bardet's eyes for the full development of the human being: this is the level of a "human city."[16]

At first glance this looks very speculative, with a specific number of families attributed to each level and multiplied by ten as one moves to the next level (with the patriarchal scale counting 5-10 families, domestic 50-150, and so on). But in fact, Bardet's "social topography," as he calls it, builds on empirical methods, and—just as with Foucault's description of the security paradigm—urbanism is considered a science which aims at defining and implementing norms according to the average level of a phenomenon in question discovered in reality itself. Accordingly, Bardet dwells on methods developed in French sociology, history and human geography since the late nineteenth century, distinguishing "community levels" by means of specific economic patterns and everyday practices, the intensity and frequency of social contacts and the distribution of urban functions. Thus, if the patriarchal scale is defined by 5 to 10 core families, it is because this is the group of people who exchange gifts and services, cele-brate as well as mourn together. The next level—the domestic level—is born from the proximity between neighbors, and groups of children playing together and housewives shopping together, who, as Bardet fantasied, meet "to exchange gossip." Finally, the parochial level encompasses between 500 and 1,500 families, and corresponds to an Anglo-American neighborhood

[16] Gaston Bardet, "Les échelons communautaires dans les agglomérations urbaines," in *Pierre sur pierre* (Paris: Editions L.C.B., 1945), 233-49.

unit, defined principally by the maximal distance children travel to and from their primary school—the school, then, becomes the urban element that replaces the parish church as the center of a community. In Bardet's account, the everyday life of a child, not yet fragmented by the Fordist rhythms of work and leisure that characterize the everyday routines of an adult, remains the preponderant criterion for defining a community level.[17] This also points at the fact that the increasing mobility and fragmentation of everyday life is the main challenge to Bardet's theory: the very challenge addressed by the Team 10, from the 1950s onwards.

Bardet's contribution was directed both against the European centralized cities (the three "monsters" of Paris, Lyon, and Marseille) and the functionalist urbanism of Le Corbusier and the CIAM. In opposition to both functionalist zoning (considered by Bardet as mechanical, abstract, and expressing capitalist exploitation) and also against the concept of the neighborhood unit (leading, according to him, to city fragmentation) the community levels would facilitate the organic coherence of social groups and would thereby allow for personal development of each and every individual. The city would change completely: rather than a concentric scheme, it would become a cluster [une grappe] of villages or parishes. His aim is thus to create proximity within small closed societies that add up to a large, but "open," society.[18]

Yet it was already at the time when Bardet published his essay on community levels that the discourse of CIAM urbanism took a self-critical turn towards pre-war discussions within the organization, employing arguments that not rarely coincided with Bardet's. To take one notable example, in his paper "The Human Scale in City Planning" (1944) José Lluís Sert revised the functionalist approach arguing for its "humanization." In doing so he questioned the pre-war enthusiasm for the machine, in turn opposing tendencies towards urban sprawl and suburbanization. Taking the number of inhabitants and the composition of functions as his starting point, Sert devised a hierarchy of social and spatial scales, ranging from the neighborhood unit, through the sub-city or township, the city proper, the metropolitan area, and the economic region. With the concept of the community complementing the functionalist triad of the "sun, air, greenery,"

[17] Ibid.
[18] See also Jean-Louis Cohen, "Entretien avec Gaston Bardet," *Architecture, Mouvement, Continuité* 44 (February 1978): 78-81, and Jean-Louis Cohen, "Gaston Bardet, un humanisme à visage urbain," *Architecture, Mouvement, Continuité* 44 (February 1978): 74-7.

such conceived urbanism aimed at the "design and support of human contacts" and "raising the cultural level" of the population [see image 5].[19]

Sert's article expressed the increasingly dominant idea in post-war CIAM of a hierarchy of spatial entities that were both to reflect and to facilitate the constitution of a community. This was conveyed by a project of a new town for 30.000 inhabitants in the Furttal valley near Zurich, launched in 1957 and developed during several years by an interdisciplinary team headed by Ernst Egli, professor of urbanism at the Swiss Federal Institute of Technology (ETH) in Zurich. The project was based on a matrix of seven levels of "human organization" combined with a list of twelve basic needs.[20] Egli underscored the role of sociology in the design by claiming that the urbanist "would be grateful if the sociologist could provide him, sociologically speaking, with a useful, spatial net of relationships in the city."[21] This vision of urbanism as realized sociology resulted in a hierarchy of social groups, starting from the individual, through the family, the neighborhood, a group of neighborhoods, a small district, a district, and up to the city itself [see image 6]. According to Henri Lefebvre, who reviewed this project in 1960, an isomorphism between social and spatial entities is assumed: "one composes the community with families like the functions of the city with elementary needs attributed to various levels."[22]

Team 10 and the urbanism of the in-between

This cursory move from Bardet, a major exponent of French "culturalist" urbanism, through the evolution of "progressivist" urbanism of Sert or Egli, suggests that notable representatives of what Françoise Choay identified in the 1960s as antithetical tendencies in twentieth century urbanism,[23] shared the fundamental assumption about the obligation of urbanism to interrelate nested hierarchies of social and spatial morphologies. The way in which this

---

[19] José Lluís Sert, "The Human Scale in City Planning," in *New Architecture and City Planning*, ed. Paul Zucker (New York: Philosophical Library, 1944), 392-412.
[20] Ernst Egli, Werner Aebli, Eduard Brühlmann, Rico Christ, and Ernst Winkler, *Die Neue Stadt: Eine Studie für das Furttal* (Zurich: Verlag Bauen & Wohnen, 1961).
[21] Ibid, 53.
[22] Henri Lefebvre, "Utopie expérimentale: Pour un nouvel urbanisme," in *Du rural à l'urbain* (Paris: Anthropos, 1970), 129-40, here 135. See also Łukasz Stanek, *Henri Lefebvre on Space: Architecture, Urban Research, and the Production of Theory* (Minneapolis: University of Minnesota Press, 2011).
[23] Françoise Choay *L'urbanisme: utopies et réalités: une anthologie* (Paris: Éditions du Seuil, 1965).

interrelationship was conceived came under fire by the third generation of the CIAM and the Team 10, which challenged the definition of a community by means of geographic isolation. Convinced that "the creation of non-arbitrary group spaces is the primary function of the planner," the members of the Team 10 introduced the concept of the "hierarchy of human associations," inspired by the "valley section" of the Scottish biologist Patrick Geddes (1909).[24] This hierarchy, which the Team 10 postulated in exchange for the Athens' Charter, was not defined by means of bounded spaces: "the principal aid to social cohesion is looseness of groups and ease of communications rather than the rigid isolation of arbitrary sections of the total community with impossibly difficult communications, which characterize both English neighborhood planning and the 'unité' concept of Le Corbusier."[25]

This was developed in the *Team 10 Primer* (1962), edited by Alison Smithson, by means of three categories used as chapter titles: "Urban infra-structure"; "Grouping of dwellings"; and "Doorstep." What at first glance suggests a hierarchy of scales—from that of the city, though a housing neighborhood, to an individual apartment—refers, rather, to three modes of defining scalar hierarchies. First, urban infra-structure (such as large-scale road system) was considered as foundational for the identity of the community, defined through movement: it is through the hierarchies of movement that various spatial and social scales are established. The question of dwelling, secondly, introduces housing as the criterion for a different scalar hierarchy—from the house, the street, the district, to the city—all defined by mobility and communication of groups of people: a modulated continuum of scales which became a major theme for the Golden Lane Deck Housing project. This discourse was based on the imagination of a traditional city ("it is the idea of street, not the reality of street, that is important")[26] and in subsequent years it was replaced, in the discussions of the Team 10, by a more abstract language about "stem," "cluster," and "cell."[27]

In contrast to two other categories from the *Primer*, the "doorstep" stands for a different type of understanding of scale, developed in the texts of Aldo van Eyck. For van Eyck, the doorstep is an in-between sphere, in

---

[24] Alison Smithson (ed.), *Team 10 Primer* (London: Standard Catalogue Co., 1962), 78.
[25] Ibid, 78. See also Volker Welter, "In-between Space and Society: On Some British Roots of Team 10's Urban Thought in the 1950s," in Dirk van den Heuvel and Max Risselada (eds.), *Team 10, 1953–81: In Search of a Utopia of the Present* (Rotterdam: Netherlands Architecture Institute, 2005), 258-63.
[26] Smithson, *Primer*, 80.
[27] Ibid, 88.

which polarities are reconciled: the individual and the collective; the outside and the inside; the unity and the diversity; the part and the whole, the large and the small; the many and the few as well as the opposition between architecture and urbanism. The failure of modern city planning, according to van Eyck, stems from its inability to deal with these "twin phenomena" as he calls them: "Failure to govern multiplicity creatively, to humanize number by means of articulation and configuration [...] has led to the curse of most new towns."[28] The role of both architecture and urbanism is to define a configuration of clearly delineated intermediary places; in other words, scales are not defined any more as bounded entities but rather as a set of in-between realms.

While much of the discourse of the Team 10 was a response to the Fordist society, this understanding of the in-between realm announces a different type of discourse about the city, one that, from the late '60s onwards, became increasingly dominant. It was marked by a proliferation of debates about "intermediary spaces," "semi-public," "semi-private," "spaces of transition," "spaces of negotiation," and "urban voids"—a vocabulary that, indeed, governs discourse about urban spaces to this day.[29] Such modulation of the in-between spaces puts to an end the fundamental dialectics which defined the social-democratic imagination of much of the modern movement as well as the architectural and urban practice of the welfare-state: the dialectics between the *Existenzminimum,* on the one hand, and, on the other, the "collective luxury" of sun, air, greenery, and social facilities which are calculated according to the density of the inhabitants within specific scalar thresholds.[30] In the course of the last thirty years this dialectics has been increasingly replaced with an architecture and urbanism charged with the task, in the words of Aldo van Eyck, to create an "interior both outside and inside."[31]

Consequently, such conceived urban space becomes increasingly modeled according to intimate links between a bedroom, a kitchen, a living room, a staircase and a garden. And thus it is not accidental that much of the critique of the Fordist city—functionalistically fractured into spaces of work, housing, leisure and transportation—was developed, during the 1960s

[28] Ibid, 100.
[29] Christian Moley, *Les abords du chez-soi, en quête d'espaces intermédiaires* (Paris: Éd. de la Villette, 2006).
[30] See also: Łukasz Stanek, "Collective Luxury: Architecture and Populism in Charles Fourier," *HUNCH* 14 (2010).
[31] Smithson, *Primer,* 104.

in France, Netherlands, the United Kingdom, and the United States from within the bounds of both sociological and ethnographic research about the domestic interior, preparing architecture for its emerging role as a mass medium of normalized images of domestic consumption.[32] From that point onwards, the domestic interior and the city have become increasingly intertwined into one urban field of production and reproduction: a set of in-between spaces whose articulation is dominated by concerns of privacy, identity, and security.

---

[32] See Stanek, *Henri Lefebvre on Space*; and the recent analysis of the work by Venturi and Scott-Brown by Reinhold Martin in his *Utopia's Ghost: Architecture and Post-modernism, Again* (Minneapolis: University of Minnesota Press, 2010), 4ff.

Staging a Milieu: Surfaces and Event Zones

Helena Mattsson

Public space is diminishing in many urban areas, and it has been claimed that this implies a major transformation of the contemporary city. At the same time, public spaces are more than ever being constructed by companies and corporations, both inside and outside institutions. Here I would like to discuss a contemporary tendency to invert the relation between the public and the institution. Internal activities and programs, once hidden within organizations, are now being made visible and put on display. Thus, the re-organization of the institution affects the relations between *public space* and the *workplace*; two categories, traditionally separated, are intertwined in a kind of double bind, where both worker and public are captured by the visibility of the other.

I will take my cues from Michel Foucault's notion of a "space of security," as well as from some oft-neglected perspectives in Jeremy Bentham's *Panopticon*. Once both society and the "apparatus of the institutions" have been rendered more efficient, new forms of control begin to replace the regime of discipline. In his short but often quoted essay, "Postscript on the Societies of Control,"[1] where he develops Foucault's ideas in a new direction, Gilles Deleuze claims that all enclosed milieus have entered into a deep crisis and are subjected to continuous reform. He concludes by remarking that he sees no future for such institutions: "Everyone knows that these institutions are finished, whatever the length of their expiration periods. It's only a matter of administering their last rites and of keeping people employed until the installation of the new forces knocking at the door."[2]

Foucault's analysis of Jeremy Bentham's Panopticon as the iconic structure of discipline is often taken to be rather limited in its focus on archi-

---

[1] Gilles Deleuze, "Postscript on the Societies of Control," trans. Martin Joughin, *October* 59 (Winter 1992).
[2] Ibid, 3.

tecture as a set of physical elements—walls, windows, doors, etc.—as if it would be those particular elements that create the disciplinary regime. Thus a fundamental critique has been voiced by those who focus on digital transformations, and instead advocate a "dataveillance critique" that highlights consumer databases, personal privacy, and other non-material mechanisms of control.[3] Instead of turning the gaze towards non-material forms of control, I would like to focus on the material institution and ask: What is in fact happening to these physical places? Is it only a matter of administering their last rites? Rather, I would argue that institutions remain operative and, moreover, are developing new forms of regulations with new implications.

To understand these new disciplining structures and their relation to control mechanisms we need to focus on the material arrangement in space, but to extend this so as to include its immaterial aspects as well. When Nigel Thrift criticizes Foucault for being too narrow in his reading of the Panopticon, he points out that one such blind spot is his lack of consideration of "affect," adding that "the obvious explanation [for this] is Foucault's concentration on power, in contradiction to desire."[4] Instead one might foreground other aspects of the Panopticon, such as Bentham's interest in construction materials (especially cast iron and glass) along with the effects they helped to create, for example, the leasehold contract and issues surrounding publicity. As is well known, Bentham made detailed explanations and drawings of different plans of the Panopticon; the built structure was supposed to have large windows and minimal walls. With the use of modern building techniques and materials, the aim was to make the construction as transparent as possible.

In this respect the Panopticon is a true modernist architecture, both in its constructive and functional organization—it is a transparent machinery producing new, more efficient and healthier subjects, and in this way rationalizing society.[5] In certain institutions, such as the school, the walls could even be reduced to sail cloth, and, as the architectural historian Robin

[3] For more on this critique, see Greg Elmer, "A Diagram of Panoptic Surveillance," *New Media Society*, Vol. 5, No. 2 (2003): 232.
[4] Nigel Thrift, "Overcome by Space : Reworking Foucault," in Jeremy W. Crampton and Stuart Elden (eds.), *Space, Knowledge and Power: Foucault and Geography* (Burlington: Ashgate, 2007), 54.
[5] Charles F. Bahmueller argues for the connection between the Panopticon plan and the welfare state, and suggests that the essence of both is to produce more well-being and more efficient subjects. See *The National Charity Company: Jeremy Bentham's Silent Revolution* (Berkeley and Los Angeles: University of California Press, 1981), 103ff. Gertrud Himmelfarb criticizes this thesis in her review of the book, in *The Journal of Modern History*, Vol. 56, No. 1 (March 1984): 139-140.

Evans points out, Bentham liked to draw attention to the similarity between the "apparent omnipresence" of the invisible governor of the Panopticon and the qualities ascribed to God.[6] This theme is elaborated in one of Bentham's last versions of the Panopticon where the inspectors were placed inside paper containers perforated by small holes, hanging from the ceiling in the center of the building like lanterns.[7] Even though Bentham's utopia for a "pauper management"[8] is his most developed project when it comes to the use of materials and light, it is also a modulation of light that makes it possible for the guard in the prison to be absent, thus installing a process of self-disciplining among the prisoners.[9]

Creating a milieu

The examples that will be scrutinized in the following, broadcasting houses and factories, can provide us with striking examples of how old forms of institutions have been re-modeled into new forms of control regulated by affect. What is common to all of them is the construction of a milieu in which audiences, or the public, are connected with employees and workers. I propose to call this milieu an *event zone*, in which the individual is trapped and controlled in a double bind: on the one hand, this zone links inside and outside, public and private, work and leisure and so on, through visible connections; on the other hand, these different milieus remain separated by material and immaterial borders.

The current transformation of the original 1932 BBC Broadcasting House (designed by Colonel G. Val Myer) from a closed monument into a transparent "platform" could serve as an example of such an institutional change.[10] The old building had a massive façade constructed by heavy stone

---

[6] Robin Evans, *The Fabrication of Virtue: English Prison Architecture, 1750–1840* (Cambridge: Cambridge University Press, 1982), 206.

[7] Ibid, 207.

[8] First published as "Outline of a Work Entitled Pauper Management Improvement," in *Annals of Agriculture*, 1798.

[9] This is pointed out by Greg Elmer in "A diagram of Panoptic Surveillance," 234.

[10] For a more developed analysis of the new transparency of the BBC, see Helena Mattsson, "The Real TV: Architecture as Social Media," in Staffan Ericson and Kristina Riegert (eds.), *Media Houses: Architecture, Media, and the Production of Centrality* (New York: Peter Lang, 2010). For other studies of the BBC building, see Staffan Ericsson, "The Interior of the Ubiquitous: Broadcasting House, London," and Kristina Riegert, "The End of the Iconic Home of Empire: Pondering the Move of the BBC World Service from Bush House," in ibid.

materials and displayed an ornamented and decorated shell facing the public. The division between the street outside and the interior of the institution was clearly marked, and the massive door at the entrance set up a threshold separating producers and consumers. Instead of being a screened-off facility for production, the new BBC-structure is open. The old building has been re-organized and extended with a new building, and together those two parts create one continuous block. By setting up a new exterior space, Langham Place, which flows into the building, an in–between milieu—or *event zone*—is created, in the form of restaurants and shops, which is also intended to house temporary events. This public space is meant to serve as a stage for the BBC, with a display of "public art" to attract audiences, as well as providing a means for live, on-stage broadcasting. This could be described as a *production of the public itself*—and, thus, of consumers.

This new public space will be enhanced through an arcade that runs through the building, providing an opportunity for the public to gain access to the interior without passing security controls. The public arcade will serve as a foyer for the Radio Theater and other public facilities such as cafés, exhibitions, and shops. In mixing functions and thus allowing the public space to encroach the space of media production, the clear demarcations between city space and interior, private and public, which had once existed, are now rendered more fluid. As a visitor, it is possible, without any preparation, to end up being part of a TV-show, a radio-interview or an artistic happening. In this floating space of multiplicities, diverse functional schemes are intertwined with multifaceted experiences. The earlier material borders, such as the thick walls that once demarcated the institution, are now replaced by event-zones that instead control the visitor by and through engagement.

In what way should the new forms of regulations and discipline be interpreted in relation to such open and "enabling" urban landscapes? There are no longer any guards, instead there is entertainment. Unlike the guard, the entertainer does not control individuals, but rather, through creating attention, has the role of capturing the public, in an open territory organized by actions. It is desire, and not restriction, that regulates the individual's movements and actions in this space. As the virtual world expands, and people spend more time in front of the computer, corporeal sensibilities, which today make architecture and the built environment central machines for subject production, become increasingly exclusive.

Even in a non-democratic country like China we can find a strategy similar to the one adopted by the BBC, namely the attempt to integrate the public by creating a milieu for amusement. The current headquarters of

China Central Television (CCTV) has separated broadcasting from production.[11] The tower dedicated to broadcasting is open to the public and works as a tourist attraction. Production, on the other hand, takes place in a closed environment of high-rise buildings surrounded by fences and guards. The new CCTV complex will contain all the facets of television production, and the site will be open to the public, with the media park envisaged as an extension of the green areas planned in the area. But while it is a site for production, the space can at any time be shut off and once more be turned into private CCTV property, so that roads formerly integrated into the urban grid become disconnected from the rest of the city. In this way we might speak of a temporary staging of a public space. According to the plans, the core of the building will be a "public loop," offering the audience a multitude of experiences.

Spaces of compensation

Architecture is often used as a surface for projections of another real space, more perfect and better arranged, that should become actualized in reality. What appears to be dismissed in our current situation is in a certain way resurrected in architecture. As mentioned earlier, it seems that public space, at the same time as it contracts, is once more re-staged through architecture, so as to reemerge inside new projects (institutions). Work and activities in the workplaces are opened up and, through architectural techniques, tuned into spectacles. This spectacle is in direct proportion to the increasing invisibility and opacity of the structural logic of production and consumption in a global economy. While local production is dependent on diffuse global networks of suppliers, making thereby the whole process of production resistant to any general analytical accounting, work and production become spectacles.

Foucault speaks about the *heterotopia of compensation*—"a space that is other, another real space, as perfect, as meticulous, as well arranged as ours is messy, ill constructed and jumbled."[12] This could be understood as a compensation for the gaps and losses in the real world. Some of the spaces discussed here could be described as heterotopias in Foucault's sense, al-

---

[11] For a more developed analysis of CCTV see Helena Mattsson, "The Real TV: Architecture as Social Media," and Sven-Olov Wallenstein "Looping Ideology: The CCTV Center in Beijing," in *Media Houses*.

[12] Michel Foucault, "Of Other Spaces," trans. Jay Miskowiec, *Diacritics*, Vol. 16, No. 1 (Spring 1986): 22-27.

though they are not necessary separated from the spaces or situations that they reflect, so that the "other space" is in fact fused together with the space that it reflects. The lack of public space in the city is compensated for by the resurrection of public space inside architectural projects, and inside privately owned and planned spaces. Here no difference between private and public appears to exist; such categories *seem* to have lost all meaning.

Publicity as a tool for creating a space of security

In his lectures at the Collège de France, from 1977–78, Foucault formulates an alternative technique to the disciplinary space: the "space of security."[13] In security, we are dealing with spaces that allow for multiple intersecting events, and even if Foucault does not explicitly elaborate this in terms of desire, we can take this to be a basic drive underlying all non-predictable events and developments in the city. Architecture indeed has a capacity to induce public affect, to stage a milieu, or—to use Foucault's terminology—a space of security, which is a space that lets things happen, contrary to the disciplinary version, where events were regulated through divisions. In this milieu the traditional institution becomes a zone of amusement where even work is perceived as entertainment.

In both of the above examples, BBC and CCTV, architecture is used to attract and capture audiences by the creation of a milieu of affects. The public must in fact be understood not as a pre-given entity, but as the name of a *technique* for staging a milieu in Foucault's sense of the concept, space as a medium for events. Programs and activities that earlier were hidden in the machinery of society are now opened up for the public. Through transparent surfaces or event zones the individual is caught up in unexpected, even though pre-staged, situations. When society and the state apparatus have been rendered more efficient, the panoptic machines must be re-built or re-organized: the tower is opened up to the public, the entertainer takes the place of the guard, and instead of monitoring the individual he now addresses the public.

---

[13] Michel Foucault, *Security, Territory, Population: Lectures at the Collège de France 1977–1978*, trans. Graham Burchill (New York: Palgrave MacMillan, 2007). Foucault understands these different form of techniques as existing in a complex relation to each other, so that "there is not a series of successive elements, the appearance of the new causing the earlier ones to disappear" (8).

Already in Jeremy Bentham's liberal Panopticon utopia, the doors were "thrown wide open to the body of the curious at large: the great open committee of the tribunal of the world."[14] Making the Panopticon open to the public was a way to guard the guards, and minimize the possible misuse of power in the prison. In this way both entertainment and surveillance can be practiced simultaneously, and a visit might, with Bentham's words, "satisfy a general curiosity which an establishment, like the prison, may naturally be expected to excite."[15] Furthermore, we must note that the Panopticon should be privately owned, and be organized as a corporation. The idea of private profit was crucial for Bentham's plan for the construction of the correction institutions: "This is the only shape which genuine and efficient humanity can take. Every system of management which has disinterestedness, pretended or real, for its foundation, is rotten at the root...."[16]

In recently constructed public event zones, like the BBC or the CCTV, forms of private ownership create new power structures through a stronger interdependency between the state, the municipality, and the corporations. The staging of public space has been outsourced to consultants, companies, and institutions. The municipality states this to be a pre-requisite for building, as in the case of Nike in New York City. At the same time we can notice a "becoming-amusement park" of the public space, in which companies, branding themselves through the creation of spaces, identify everyone as a possible consumer. These two tendencies seem to overlap in an efficient way. Public space remains as a staging of private space, which satisfies the public authorities, at the same as this opens up the possibility of producing consumers and audiences as an integral part of a staged milieu.

In 1998 Nicolas Bourriaud coined the expression "esthétique relationnelle," in the same year Kevin Kelly published *New Rules for the New Economy*, and the year after Pine and Gilmore's *Experience Economy* was released.[17] These books established, if not individually then at least as a cluster, a close link between art and marketing. In this "new economy" consumers must be made active and entertained, through the production of

---

[14] Jeremy Bentham, *Panopticon or the Inspection House* (Dublin: 1791), 33.

[15] Ibid.

[16] Ibid.

[17] Nicolas Bourriaud, *Esthétique relationnelle* (Dijon: Presses du Réel, 1998); Kevin Kelly, *New Rules for the New Economy: 10 Ways the Network Economy is Changing Everything* (London: Fourth Estate, 1998); Joseph B. Pine and James H. Gilmore, *The Experience Economy: Work is Theatre & Every Business a Stage* (Boston, Mass.: Harvard Business School, 1999).

experiences: it was an "'emotional or passionate economy,' which also meant highlighting aestheticization and performative qualities."[18]

Experience work

As mentioned earlier, the relation between the public and workers is reformulated in these transparent re-organized institutions. In both Taylorism and Fordism the link between work and product was severed and the product was transformed into a commodity with supernatural characteristics, a disruption that we may understand through the Marxian concepts of alienation and commodity fetishism. Displaying the commodity as a magic object in a shop window was meant to enhance its character as a fetish. The "dirty work" that preceded the finished product took place in factories closed to the consumer, often located at the outskirts of urban space. Taylor's "army of gorillas" was hidden from the public, together with the unfinished products, for both political and economic reasons.

Today, however, the scene is inverted, though again for reasons of both political and economic expediency. The factories are located inside the cities; the so-called site of production is open to the public and the commodity is on display throughout the production process. If, before, it was *production* that took place in Taylor-like factories, then today, what in the modern "factory" takes place is the *assemblage*. In an experience economy the dividing line between production and consumption is redrawn; it is *not* obliterated, even if it may seem so. In late capitalism the production process is global, and if it was the walls of factories that once made production invisible, then today it is territorial distances that divide the consumer from the substantial part of the production process.

The *Transparent Factory* (*Die gläserne Manufaktur*) in Dresden, designed by Henn Architekten, is a modern factory located in the city and open for the public. It is presented as "the only place in the world to turn production into a real experience," a place where a "new transparency" functions as the surface of communication: "We stage what usually takes place behind closed doors as a place of communication and exchange."[19] The work on display is clean, almost clinical, and the workers all wear white

---

[18] Orvar Löfgren and Robert Willim (eds.), *Magic, Culture, and the New Economy* (Oxford: Berg, 2003), 2.
[19] See http://www.glaesernemanufaktur.de/gmd.jsp?dok=&lang=&docid=&ap. Accessed 2009-06-08.

overalls; the factory has been aestheticized to the point of appearing like an art gallery, and production is akin to art production. A similar organization could be found in the BMW factory in Leipzig, designed by Zaha Hadid. This is a modern factory located in the city. In the factory, three central segments of the production chain—body shop, paint shop, assembly line— are opened up to each other, "showing each one of the workers how their roles at BMW are interrelated." BMW also offers a public tour through the various stations: "Take a look behind the scenes and experience live how a BMW is built."[20]

Conclusion

This essay shows how contemporary architecture produces new types of public spaces and workplaces as compensations for gaps and losses in our world. In these spaces of compensation, internal activities and work are displayed and are turned into a spectacle. This also points towards a shift in the technique of controlling spaces from surveillance to relational engagement. These tendencies in architecture, and in society at large, are here discussed specifically in relation to the media institution and the factory. Before, what was emblematic of public space was that it was both open and empty, like Haussman's Paris; what such open and empty space made possible was the control of large territories. Such space however remained a potential stage for protests, demonstrations, and revolutions. Today, the staged event zone is a space of security, where the role of the overseer has become vacant; event zones are instead spaces in which everyone is always part of a relation, in a staged milieu of affects.

[20] See www.bmw–werkleipzig.de/leipzig/deutsch/lowband/com/en/index.html. Accessed 2009-06-08.

Neuropower: Is Resistance Fertile?

Warren Neidich

During long periods of history, the mode of human sense perception changes with humanity's entire mode of existence. The manner in which human sense perception is organized, the medium in which it is accomplished, is determined not only by nature but by historical circumstances as well. The fifth century, with its great shifts of population, saw the birth of the late Roman art industry and the Vienna Genesis, and there developed not only an art different from that of antiquity but also a new kind of perception. [---] They did not attempt – and, perhaps, saw no way – to show the social transformations expressed by these changes of perception. The conditions for an analogous insight are more favorable in the present. And if changes in the medium of contemporary perception can be comprehended as decay of the aura, it is possible to show its social causes.

Walter Benjamin, *Illuminations*

The three facets of neuropower

Neuropower constitutes the new focus of power to administer difference in order to sculpt a people.[1] It consists of three key concepts. First and foremost it acts upon the neural plastic potential of the brain in a living present, especially during what are referred to as the critical periods of development, all the time being guided by the desire to produce a conscripted and en-

---

[1] Gilles Deleuze, "Postscript on the Societies of Control," *October* 59 (Winter 1992): 3-7. The passage from the disciplinary society to the society of control and noo-politics, that is to say, the administration in the closed and wide-open spaces, previously focused on the condition of the individual and the dividual in relation to the past and the present. They described the focus of power as that which organized the interruptions and undulation of flows of time and space in the disciplinary society and society of control, respectively, in the context of a "present condition of the now," even if, for instance, as in the society of control Deleuze suggests future kinds of gadgets of control, such as an "electronic card that raises a given barrier."

rolled individual of the future. Critical periods are temporal windows in which the nervous system is especially sensitive to the effects of the environment mediated for the most part by parental influences early in life through what the great Russian Psychologist L.S. Vygotsky called internalization or the internal reconstruction of a formerly external activity.[2] The acquisition of language is internally reconstructed and is coupled to a process called epigenesis in which even local cultural influences can play an important role in sculpting the pluripotential of the brain. Epigenesis is defined as the means through which the unfolding of the genetically prescribed formation of the brain is altered by its experiences with the environment, whether that be the milieu of the brain itself or the world. At one time, when man lived in nature it was nature that had provided the experiences to alter the brain. Today, as more and more people move to the designed spaces of the city, it is culture. When one considers brain function in this context, the term neural plasticity is used. Neural plasticity delineates the means through which the components of the brain—that is, its neurons, their axons, dendrites, synapses and neural networks (refered to as its firmware)—in addition to its dynamic signatures, like temporal binding, which allow distant parts of the brain to communicate, are modified by experience. For instance, the immature brain has the capability of learning over 6.700 different language variations, even if it chooses to learn only one or a few. The Japanese child growing up in London can learn English perfectly, without any trace of an accent, as can the English child growing up in Tokyo.

> Human infants have special cognitive abilities that are built for exactly this cultural variation. For example, in the realm of vowel sounds, infants of just six months have been shown to restructure their auditory space according to the local language; the space becomes systematically and irreversibly distorted [...] The end result is a range of spectacular biases in our auditory perception, which make adults unable to even hear the difference between sounds that are fundamentally distinct in some other language.[3]

---

[2] L.S. Vygotsky, *Mind in Society* (Cambridge, Mass.: Harvard University Press, 1978).
[3] Stephen C. Levinson, "Introduction: The Evolution of Culture in a Microcosm," in Stephen C. Levinson and Pierre Jaisson (eds.), *Evolution and Culture* (Cambridge, Mass.: MIT, 2006), 14. Note that the words "cultural variation" are used to refer to language learning.

Secondly it redirects the armamentarium of power from a focus upon distributions of sensations, as elaborated by Jacques Rancière,[4] with its concomitant forms of bottom-up processing—in which abstract concepts are built from concrete sensation—to one focusing upon top-down processing: abstract concepts centered in the forebrain and pre-frontal cortex modulate future actions and behaviors by affecting the downstream sensorial and perceptual systems, to which the brain is connected. These abstract concepts are formed in the working memory.

Today, it can be advanced that mechanisms or apparatuses of power have increasingly found ways to intervene in the working memory, doing so through the rearrangement of its contents. The working memory refers to memories held briefly in the mind, making possible the accomplishment of a particular task in the future. Important in this regard are the conditions of new forms of machinic intelligence and competence in the age of immaterial labor, alongside a notion of general intelligence prescribed by tertiary economies, in which worker choice and participation in decision making play an increasingly crucial role. The frontal lobe is essential, for instance, in what is referred to as free-choice situations, according to which one must decide how to interpret an ambiguous situation. In this regard the new focus of power is not only on the false reproduction of the past—analogous to manipulating an archive; the effects of power have moved to the reconstitution of the working memory, elaborated by the forebrain in the making of a plan.

Can the new burgeoning fields of consumer neuroscience and neuro-economics provide the methodology to influence these decisions making patterns through interventions in the working memory itself? Is the recent success of the film *Inception* (2010) a response to our societies' collective anxieties about the possibility of memory espionage? The frontal lobes, as opposed to the senses, are the new focus of power and, *mutatis mutandi*, are thus to constitute a new object for the theory of power. While acknowledging the importance of some of the theories of Jacques Rancière, some of which are built upon here—specifically his ideas surrounding the distribution of the sensible, its policing and the artist's role in rearranging it—this essay nonetheless notes the diminished role that such an analytic may play in the future. This article calls for the development of a designed post-phenomenology, in which sensation and perception are bypassed. It advances that it is the organization of memory during the production of a plan—and

---

[4] See Jacques Rancière, *The Politics of Aesthetics* (London: Continuum, 2006).

not straightforwardly memory itself—that constitutes the new site of administration.

Thirdly, I would like to suggest that neuropower is the latest stage of an ontogenic process beginning with the disciplinary society, as outlined by Michel Foucault, followed by the society of control, as developed by Gilles Deleuze, and proceeding onwards towards Marizio Lazzarato's noo-politics: "Noo-politics the ensemble of techniques of control) is exercised on the brain. It involves above all attention, and is aimed at the control of memory and its virtual power."[5]

Each epoch as it is defined by, for instance, new forms of social, political, economic, psychological and technological relations, requires new forms of dispositifs to administer the people. This ontogenic structure is a response to those conditions. Such an account is, however, not to be interpreted as crudely positivist and linear; on the contrary, the process is full of bush-wacking and backtracking. There are examples showing the extent to which the disciplinary society is still important today, as well as the discovery of traces of neuropower in the past. In the new information economy—characterized as it is by semio-capitalism—in which the production of objects has been superseded by the production of psychic effects and new powerful tools, (such as software agents, which trace our choices and calibrate our desire) the ability of neuropower to map institutional paradigms upon the materiality of the wet, mutable organic surface of the brain itself is being realized. New labor, as it too journeys ever closer to becoming a perfor-mance—such that praxis and poetics merge—does in fact leave a trace.

Neuropower distinguishes itself from noopolitics in two important ways. First, it is not about the modulation of the attentive networks in the real present cultural milieu, but is instead about the rerouting of the long term memories into working memory where decisions are made for an active moving body projected into the future. This is the key to its link to the performative conditions of labor in the new economy. The machinic intelligence is not in the apparatuses of production as they once existed in the assembly line of factories, but are rather installed within us as machines in cognitive labor. Seondly, neuropower is not about the production of a real object, but is exerted through a modification in the neurosynaptologics of the brain. In cognitive capitalism, neuropower works to produce changes in the material logics of the brain by affecting the brain's neurons and

[5] Maurizio Lazzarato, "Life and the Living in the Societies of Control," in Martin Fuglsang and Bent Meier Sorensen (eds.), *Deleuze and the Social* (Edinburgh: Edinburgh University Press, 2006), 186.

synapses, its so called firmware as well as its dynamic properties such as the properties of binding and reentry.

The present text does not afford me the opportunity or space to expound on the variety of political outcomes of neuropower; what I would like to do instead is to elucidate some of the above concerns through an explanation of the other side of neuropower. Similar to what Michael Hardt and Antonio Negri have contributed in their complexification of biopower, we must also consider that there exists another side to neuropower.[6] The role of art production as a means to counterbalance and challenge this power of the sovereign in the age of neoliberal global capitalism—especially in the latter's transitions into neoliberal cognitive capitalism, in which the labor of thought itself provides, on a global scale, the new territory for capitalistic adventurism—will form the subtext to what follows.

Through both its direct and indirect effect on the cultural field, by first mutating the distribution of built space and recently through rerouting its memory and attention, artistic practice can activate the pluripotentiality of neural plasticity. In its most utopian guise it can emancipate the virtual contingencies locked up in the pluripotentiality of the pre-individual, itself a result of the tremendous variation of the neurobiologic substrate, sculpting inter-subjective difference and heterogeneity. For my purposes here I shall quickly elucidate the form of this emancipation through an exposition of the way that noise music has influenced the tastes of a generalized contemporary music appreciation. This shall be undertaken with the use of John Cage's now famous *4' 33."*

Art power: Resistance is fertile

[...] Deleuze describes the brain as a "relatively undifferentiated mass" in which circuits "aren't there to begin with"; for this reason, "[c]reating new circuits in art means creating them in the brain too." The cinema does more than create circuits, though, because, like a brain, it consists in a complexity of images, imbricated and folded into so many lobes, connected by so many circuits. While the cinema can simply reiterate the facile circuits of the brain, "appealing to arbitrary violence and feeble eroticism," it can also jump those old grooves, emancipating us from the

[6] Michael Hardt and Antonio Negri, *Empire* (Cambridge, Mass.: Harvard University Press, 2000).

137

typical image-rhythms [...] opening us to a "thought that stands outside subjectivity."[7]

"Cultural Creatives"—in all their many forms as visual artists, poets, dancers, musicians, cinematographers, and so on—are able to play a role in the production of resistant cultural regimes. Such practices have important implications for thinking the mechanisms through which the fruits of artistic labor might compete for the brain-mind's attention, leading thereby to reactions and effects in the molding of the neural plastic potential. The power of art, in its most utopian sense, is to create or recognize externalities in cultural milieus as a way to release a cultural potential lingering in the "below the surface substrata" of meanings by bringing them forth, creating disparate and competitive networks that can first couple to, and then effectuate within, the brain's neural potential to become something other. Artists using their own materials, practices, histories, critiques, spaces, and apparatuses, can create alternative distributions—or redistributions—of sensibility, calling out to different populations of neurons and neural maps, potentially producing different neurobiological architectures. Some examples are necessary to make this tangible.

Think here for a moment about the relationship between Mozart's Sonata for Two Pianos in D Major—associated with producing the "Mozart effect"—and that of noise, free music or improvisation. In 1993, Gordon Shaw and a graduate student, Frances Rausher, showed that listening to the first ten minutes of this composition produced an increased ability for spatio-temporal reasoning.[8] He later concludes that the "symmetry operations that we are born with and that are enhanced through experience form the basis of higher brain function." Finally, "[p]erhaps the cortex's response to music is the Rosetta Stone for the code or internal language of higher brain function."[9] Even so, Shaw and company forget an important consideration: we still do not know how audiences first responded to this music. Maybe instead of music it initially sounded like noise. Perhaps the first audiences who listened to this work by Mozart responded in a similar way to how audiences responded to Beethoven's Fifth Symphony for the first time:

---

[7] Gregory Flaxman, "Introduction," in Flaxman (ed.), *The Brain is the Screen: Deleuze and the Philosophy of Cinema* (Minneapolis: University of Minnesota Press, 2000), 40.
[8] Gordon L. Shaw, *Keeping Mozart in Mind* (San Diego: Elsevier Academic Press, 2000), xxii.
[9] Ibid, 108.

As chronicled in Nikolas Slonimsky's perversely wonderful Lexicon of Musical Invective, even the most comfortable and cherished staples of our current repertoire, including Brahms, Chopin, Debussy and Tchaikovsky, had been condemned by contemporary esthetes in the very same way. Even Beethoven's Fifth Symphony, now the most popular classical work of all, was damned as "odious meowing"—and not music—decades after its premiere.[10]

Like those modernist observers, discussed by Fredric Jameson, who experience the postmodern space of the Bonaventure Hotel,[11] or the scandalous reception of Marcel Duchamp's Fountain (1917) in the exhibition Society of Independent Artists of the same year, earlier audiences listening to Beethoven's Fifth Symphony for the first time had not developed the perceptual habits to understand and integrate its rhythms and melodies. These artworks were sublime, because they went beyond the cognitive capabilities of neurobiologic apparatuses otherwise used to make sense of them.

But what does this have to say about noise or free music or improvesation? Rather then enlisting circuits already on hand or parasitizing already existing cerebral rhythms—noise and its bedfellows—both improvisation and free music operate, in fact, through their attempt to delink from already present patterns, creating instead resistances and emancipatory gestures. Anthony Isles, quoting Edwin Prevost, focuses on the crucial condition of improvisation and free music with particular attention to leading jazz musicians, such as Ornette Coleman. Examining how they come into being and how they are made, he notes that instead of practising a written score and matching it, "musicians train, developing their musical capacities through a process of 'de-skilling' and 're-skilling.' What these musicians are developing... [is] the ability and attention necessary to be able to respond to their co-players, to a situation and to an evolving musical time/space."[12] Each instrument plays its own score adapted to its own proclivities and idiosyncrasies. This idea of learning to pay attention to a set of gestures occurring in time—an anatomy of signs in a confined social space in which nothing is certain—produces ruptures and asynchronies.

How different, however, is the following quote to the views voiced by Gordon L. Shaw, which we encountered above: "And this musical space

---

[10] Peter Gutmann, "The Sounds of Silence," Classical Notes: http://www.classicalnotes.net/columns/silence.html.
[11] Fredric Jameson, Postmodernism, or, The Cultural Logic of Late Capitalism (Durham: Duke University Press, 1990).
[12] Anthony Isles, "Introduction: Noise and Capitalism," Kritika 02 (2009): 19.

relates to another musical time, freed from the score and freed from repetition, by neither having a set time nor tempo allotted, improvised music breaks with linear cumulative time and narrative historicization."[13] One might then ask the question: how does noise and improvisation become sensible? Referring to Csaba Toth in the same collection of texts, Isles refers to noise "as the other side of music and everything outside the discipline, literally encompass[ing] what hasn't been discovered as music yet."[14]

What was it like for an audience to first hear a John Cage performance *4'33* (1952)? *4'33* (pronounced "Four minutes, thirty-three seconds," or, as the composer himself referred to it, "Four, thirty-three") is a three-movement composition by American avant-garde composer John Cage. It was composed in 1952 for any instrument (or combination of instruments), and the score instructs the performer not to play the instrument during the entire duration of the piece, that is, throughout its three movements. For those not familiar with this work a description of its first performance by pianist David Tudor will lay the framework. First setting himself at the piano he then opened the keyboard lid and sat silently for thirty seconds. He then closed the lid and the quickly reopened it. There he sat motionless for a full two minutes and twenty-three seconds. He then closed and opened the lid one more time, sitting silently for one minute and forty seconds. Finally he closed the lid one final time and walked off stage.[15] One can find another version of the work on Youtube in which the piano is originally open and where Tudor rests a pocket watch on the lid of the piano to accurately monitor the time. Although commonly perceived as "four minutes thirty-three seconds of silence," the piece actually consists of the ambient sounds of the environment that each listener hears while it is being performed and the continued sense of unease directly following. The piece pushes each of the listeners outside his or her presumed concert space to sample their own combination of ambient sounds. Noises such as a pencil dropping, the breathing and coughing of others, one's own heartbeat as a result of one's own intimidation, a baby's cry, all become the score of an internalized and individually created composition. More importantly, this work follows Cage's more general investigation into time. By stripping the music of its musical score and laying bare its temporal underbelly, this work

---

[13] Ibid.
[14] Ibid.
[15] James Pritchett, "What Silence Taught John Cage: The Story of 4'33," in Julia Robinson (ed.). *The Anarchy of Silence: John Cage and Experimental Art* (Barcelona: Museu d'Art Contemporani de Barcelona, 2009), 167.

conflates time. Time is stretched and without its musical bearing the audiences appreciation of time is disrupted.

As early as 1937, in his now famous essay, "The Future of Music: Credo," Cage laid out some important considerations about the reception of noise. "Wherever we are, what we hear is mostly noise. When we ignore it, it disturbs us. When we listen to it, we find it fascinating."[16] Listening to a hardcore noise band in a venue like, for example, Staalplaat, in Berlin's Neukölln district, or at Jabberjaw, in Los Angeles, is for some a revelation and for others a cacophony. For others still, who are willing to linger there, a learning curve is embarked upon, as one's initial fascination with its dissonant barrage of totally nonsensical sound transitions become understandable and, indeed, pleasurable. According to Gyorgy Buzsáki, "what makes music fundamentally different from [white noise] for the observer is that music has temporal patterns that are tuned to the brain's ability to detect them because it is another brain that generates these patterns."[17] But noise as well as free music and improvisation are not sensible for everyone, even though another human brain has made it. For some what is noise will always remain so. But for others a form of adaptation does seem to occur. Are there differences between people as to their underlying cerebral circuitry and the degree to which that circuitry is modifiable? We all know older people who are very open to new things and trends, and who like nothing better than to hang out with teenagers than their own age group. Are these individuals part of a sub population who have a more supple and adaptive nervous system, one which thrives on a multiplicity of connections? Moreover, do these changing musical tastes imply more flexible and dynamic organizations, linked to unabated neural plasticity which might accommodate continued dynamic reorganizations into later life?

The appreciation, in its day, of noise and improvisation is at first localized to a limited and select population. Nonetheless, today this population has grown, with noise gaining wider recognition in mass music culture. Individuals pay money to see bands perform, they visit the venues where such performances can be found, buying and exchanging CDs or MP3-audio by their favorite artists, even though noise music remains conspicuously absent on both popular mainstream radio stations and MTV. Certain artists like John Wiese, in his recent album *Circle Snare*, are breaking this pattern and adapting noise, mixing it with punk to engage

[16] John Cage, "The Future of Music: Credo" (1937), in Cage, *Silence: Lectures and Writing* (Middletown, Conn.: Wesleyan University Press, 1961), 3.
[17] Gyorgy Buzsáki, *Rhythms of the Brain* (Oxford: Oxford University Press, 2006).

mainstream audiences.[18] Perhaps, more than simply a form of resistant experience, noise coheres around a population of brains whose perceptual habits have been formed according to a different perceptual logic, one based on an immanent field of dissonant patterns, which linger in the pluri-potential cultural field, as disjointed externalities orbiting small foci of meaning, but which have yet to join the contemporary cultural zeitgeist. Just as the brain uses miniscule portions of its temporal coding potential, culture's similarly underutilized potentiality is also the reason of its con-tinual experimentation at the margins of temporal experience.

Perhaps those who are the vanguard and thus the first to appreciate noise music are a group of individuals who favor dissonant and distressed aesthetics, like those marching to a different drummer, who prefer to cross a grassy knoll diagonally rather then follow the man-made stone pathway. Or maybe our culture has itself tuned its pattern recognition capabilities to the images and sounds of interactive medias, photographic-video hybrid apparatuses, which create typologies of topologies of disconnected patterns produced by images of incomplete bodies appropriated by the fashion industry to capture a younger generation's attention, as they are assembled on billboards framing public spaces. Such patterns are implicitly activated in, for example, the slow motion, uncoordinated falling of a recently checked hockey player—replayed over and over again on cable TV screens or monitors at sport bars—and, to offer a further example, in the particulate diffusion of spectacular light seen in the explosion of a building videotaped, which is then edited in After Effects CS-5 as action, stop action, repackaged as a QuickTime movie downloadable on YouTube, a video-clip which can even be played in reverse! On the other hand, home video programs on laptops, such as Final Cut Pro and iMovie allow anyone to be a filmmaker. Everyone is an artist, since new technologies make once difficult skills easier and more widely available. Most radical filmmaking techniques and gestures, like the montaged effects found in such movies as Dziga Vertov's *Kino-Eye* (1924), are commonplace motifs of MTV-type music videos made by amateurs found on YouTube as well as those that are incorporated into more corporate structures like the special effects and fast feed forward editing found on ESPN or the foregrounding of trucage and special effects in movies like *Time Code* (2000), in which the screen is divided in four, so as to depict different stories unfolding simultaneously, or, even, in *Inception* (2010) in which special effects create the look and feel of a video game.

[18] Discussion with Andrew Berardini, Los Angeles, 2010.

Special effects have overwhelmed other aspects of film and TV, such as plot and character, driving viewers into movie theaters as the tremendous success of *Avatar* (2009) and *Inception* (2010) would suggest.

These methodologies are directed towards a new generation of viewers who have incorporated the resulting new temporalities of the fast cut and reverse motion of the moving image into their cognitive regimes. In today's image-based culture, knowledge of these grammars of image-regimes is essential for knowing what's new, in and cool. In advertisements for products this is the new language of collage, where fast cut is indexical for youth culture and as such it participates in the avant-garde of mass consumerism. What is most important here is the way that these images capture the attention of a specific generation of subjects whose brains have been sculpted by these novel cultural landscapes. Brains cultivated in semio-capitalistic environments primed for what Paul Virilio has called phatic signifiers. In our present day world these phatic signifiers have been bound together as branded networks of phatic signfiers, which couple to similary bound global neural networks—networks that are connected throughout the cerebral cortex to the brainstem pleasure centers, in the brain. In fact these shared neurobiologic conditions produce the reification of reproducing our tastes, and these techniques of mass consumerism invent the new criteria by which to judge a new product. This knowledge is essential as it is neural selective or constructive and might even lead to a form of sexual selection.[19] If you are hip to new fashions, and perfumes, which are signified by these video styles, you may be more popular which, in turn, might lead to gaining advantage in mate selection. If cool girls or guys with this same knowledge and taste is what you are after! Such cognitive regimes constitute what Pierre Bourdieu refers to as habitus: a unique synthesis of one's genetic endowment, circumstances of birth and upbringing, and subjective experience of the social and cultural environment in which one has grown up. Are these then the new dynamic cultural signifiers determined by Holly-wood and Madison Avenue as the attention attractors for a new generation? Perhaps it is an anaesthetics of decay and destabilization that is now drifting through a population of psychic vampires hungry for new forms of sensuality and entertainment, but which in the end might create new systems of neural networks that, in their totally combined conditional feedback on self-reflection, are productive of new conditions for thought.

[19] See Gerald Edelman, *The Remembered Present* (New York: Basic Books, 1989), and Steven R. Quartz and Terrence J. Sejnowski, "The Neural Basis of Cognitive Development: A Constructivist Manifesto," *Behavioral and Brain Sciences* 20 (1997): 537-596.

The quote from Walter Benjamin's *Illuminations* that I initiated this essay with goes to the very heart of the discussion explored here. How does human sense perception change with humanity's entire mode of existence? Is human sense perception and cognition linked to changes occurring in social, political, psychological, spiritual, and economic relations, which inflect themselves through aesthetic objects, non-objects, performances, spaces, non-spaces, and which together form the semio-linguistic and cultural landscape? Such a landscape embodies those very material historical conditions, which were once responsible for its becoming, and which are subsequently coupled to various material and immaterial neurobiological relations and its mental productions—like synaptic stabilizations and prunings as well as dynamic mappings, which effect the operations of our perceptual-cognitive apparatuses. If the fifth century, with its great shifts of population, produced the birth of both the late Roman art industry and the Vienna Genesis, promoting not only an art different from that of antiquity but also a new kind of perception, then what of our own epoch as it leaps through the hoops of modernist extensive linear productivity of the assembly line into a post post-modern condition of intensive networks and non-linear on-time productivity of on-line prosuming and crowdsourcing? Antonio Negri sums this up in the following statement:

> We can no longer interpret these according to [to the classic labor theory of value that measures work according] the time employed in production. Cognitive work is not measurable in those terms; it is even characterized by its immensurability, its excess. A productive relation links cognitive work to the time of life. It is nourished by life as much as it modifies it in return, and its products are those of freedom and imagination. [---] Of course, work still remains at the center of the entire process of production [...] but its definition cannot be reduced to a purely material or labor dimension. This constitutes the first element of the caesura between modern and the Postmodern.[20]

What kind of new perceptual capabilities might this caesura engender? I wager that the new theoretical approaches—for instance, like this idea of neuropower, linked as it is to semiocapitalism and cognitive capitalism—might provide the epistemic apparatuses to engage with these questions, so as to think them anew. Walter Benjamin's intuitions are just as true today as they were then: "The conditions for an analogous insight are more favorable in the present."

[20] Antonio Negri, *The Porcelain Workshop* (Los Angeles: Semiotext(e), 2008), 20.

144

Foucault and Lacan: Who is Master?

Cecilia Sjöholm

Lacan's desire

"The master breaks the silence with anything—with a sarcastic remark, with a kick-start. That is how a Buddhist master conducts his search for meaning, according to the technique of *zen*. It beseeches students to find out for themselves the answer to their own questions. The master does not teach *ex cathedra* a readymade science; he supplies an answer when the students are on the verge of finding it."[1]

The above quote is taken from Lacan's introduction to his first seminar on *Freud's Papers on Technique*. It frames the psychoanalytic question so that it becomes a question of the master, and, moreover, becomes a question of technique. Nonetheless, this framing of the question leaves the reader with her own question: what kind of question is it that the student will ask? The quote above implies that the very technique of psychoanalysis is such that it rests on the subject of psychoanalysis asking the right sort of questions, rather than forwarding the right kind of answers. The question: "who am I?" or "what am I?" will be left unanswered, and replaced by "who is talking?"

To Hannah Arendt, the philosophical question of the Who— "who am I? Who is he"—must be substituted for the metaphysical question of the What, in order to avoid essentialism. There is no human essence in the sense of what we are. We can only think in terms of who we are. If we are to believe it is possible to have full knowledge of what man is, we must imagine a God that sees everything. It is, on the contrary, not possible to turn the question of the Who into a metaphysical question. The question of

[1] Jacques Lacan, *Freud's Papers on Technique, 1953–1954* (London: Norton, 1991), 1.

the Who points to man as a singular being, leaving metaphysical desire unanswered. This desire, however, is impossible to fully extinguish. The desire to get to know the human essence is part of human life, although such a desire can never be satiated. The Who and the What follow each other, in the condition where "I have become a question to myself."[2]

Let us pursue an examination of the question of the Who in conjunction with a reading of Foucault's concept of the technologies of the self. It may appear that Foucault's project aims towards the realization of a greater measure of freedom for the self; the technologies of the self being a quest for the possibilities that one can have in relation to the norms of society. In other words: what are my limits, what are my possibilities, within the normative framework by which my desire and my knowledge are shaped. Foucault considers the formation of the self as a striving towards knowledge. But the question of the Foucauldian subject—just as in the case of Lacan—is not "who or what am I?" If the subject were a product of technologies that shape his truth, rather than a readymade science, then the Foucauldian question would be, as argued by Judith Butler, "what can I become?"[3]

Although Foucault is a stern critic of psychoanalysis, one must note that the Lacanian focus on questions of technique and method, addressing the formation of the subject—rather than the truth of the subject—is not entirely inconsistent with a Foucauldian notion of the self. In 1988, the journal *Topoi* published an issue on the question: "Who comes after the subject?" In an interview with Jacques Derrida, held by Jean-Luc Nancy, Derrida reverses the question, by asking: who comes before the subject? The question of the subject is always placed in conjunction with the question of the Who, which in turn implies a form of submission under a law.[4] Thus, Lacan has certainly not terminated the subject, he has only displaced it from consciousness to the unconscious. As soon as we ask the question of the Who in conjunction with the subject, we point to a certain stricture that turns the subject into a singularity, responding to a universal constraint. This is certainly the case in both Lacan and Foucault: the only possible freedom is a freedom that always comes at the price of a certain submission. It is, if not an adaptation to a norm, then at least a response to a structure that is always in place. The notion of technique, as used by both Foucault

[2] A quote that Hannah Arendt associates with Augustine as well as with Paul; see *The Life of the Mind* (New York: Harvest, 1978), 65, 85.
[3] Judith Butler, *Giving an Account of Oneself* (New York: Fordham University Press, 2005), 72.
[4] *Topoi* 7 (1988): 113-121.

and Lacan, implies that the subject can in fact be re-structured, if not modeled according to certain goals. It is part of a terminology that implies a kind of opening, a certain freedom within a given structure.

To Lacan, the self is irrevocably split, and the question of truth is always placed within the framework of the unconscious. Even more importantly, the question posed by the subject is displaced by the question of the Other, as is implied by the quote on the technique of *zen*. The question of the self is tied to the master. However, the quest for identity can never be the goal for the psychoanalytic project. There can be no master answering the pledge of the patient: instead the enunciations must be brought back to the patient, and play within the relation of transference between analyst and analysand that the psychoanalytic setting gives rise to.

The phenomenon of transference is central to Freud's papers on technique published together in 1918—six papers that were originally meant to give a systematic account of the psychoanalytic technique. These papers were: "The Handling of Dream-Interpretation in Psychoanalysis"; "The Dynamics of Transference"; "Recommendations to Physicians Practicing Psychoanalysis"; "On Beginning the Treatment"; "Remembering, repeating and working-through," and "Observations on Transference-Love." However, as James Strachey points out in his introduction to the *Standard Edition*, the writings on the technique of psychoanalysis do not form a system, and, indeed, there seems to have been a great deal of reluctance on the part of Freud to complete his work.[5] There is, however, a reason for his reluctance, and one integral to the work of psychoanalysis as such. Reading Freud's papers, one must conclude that technique is a question concerning the clinic, and not theory. Therefore, the question of technique could only be formulated in conjunction with the experience of analysis.

In his own reading of the papers on technique, Lacan quickly leaves the bibliographical details behind. His readings of these papers are soon displaced by observations about the general methodological questions preoccupying Freud. Lacan observes a specific form of development: Freud becomes increasingly aware of his role as master, an awareness causing a great deal of discomfort. Thus, in Lacan's view, the papers on technique mirror a certain development of Freud's thought, based on his experience as an analyst. Famously, Lacan concludes his reading of Freud's papers on technique by advancing that new theoretical tools are necessary in the

---

[5] See the introduction by Strachey in *The Standard Edition of the Complete Psychological Works of Sigmund Freud*, Volume XII (1911–1913); *The Case of Schreber, Papers on Technique and Other Works* (London: Hogarth, 1958), 83-88.

clinic, tools that will for the subject make possible the unraveling of the question of desire, which remains the focus of analysis. In the seminar, Lacan breaks with the analysis of the ego, which had up to that point been dominant in psychoanalytic discourse. Lacan pursues the quest for a form of psychoanalysis that will reformulate its questions. Rather than being: "who am I?"—the question raised by the ego—the focus of psychoanalysis will be to target the desire of the subject. To the early Lacan—or the Lacan that held the seminar on the technique of Freud—the question of the subject must begin with the recognition that psychoanalysis is not a system, but a technique. Freud began the history of psychoanalysis with an analysis of himself. The psychoanalytic situation entails a split of the subject. Through the process of the cure, the question of the subject appears not to concern the self, but the other: *You are this.* The answer to this question, however, can only be placed as an ideal, and never fully appears.[6] The most important questions of psychoanalysis are those that concern transference and countertransference. In his seminar on technique, Lacan rephrases the question of the self so that it becomes a question of the Who: "who, then, is it who, beyond the ego, seeks recognition?"[7]

Lacanian analysis always places desire at the center. In one of Lacan's later seminars, the famous formula of desire becomes "Che vuoi?"[8] Entailed here is a distinct displacement of the subject; the subject is repositioned in relation to a Master who underpins its fantasies. "What do you want?" is the question that forms the focus of fantasy towards which psychoanalysis is directed. This is also why transference and countertransference are the most important aspects of the technique of psychoanalysis: in discovering these phenomena, Freud implies that the subject is a function of desire, rather than a function of knowledge.

Foucault's technologies

Is there, then, any relation at all between the question of technique such as it was formulated by Freud and Lacan's seminar on the one hand, and Foucault's technologies of the self on the other? Lacan's seminar on Freud's technique was offered at a time when Foucault was interested in psycho-

---

[6] Lacan, *Freud's Papers*, 3.
[7] Ibid, 51.
[8] Jacques Lacan, "The Subversion of the Subject and the Dialectic of Desire in the Freudian Unconscious," in *Écrits*, trans. Bruce Fink (New York: Norton, 2004), 312.

analysis. He visited the seminar of Lacan at Sainte Anne, and must have come in contact with the Freudian notion of technique in this way.[9] Lacan, in his seminar, equates the question of psychoanalysis with the question of technique. In his famous essay on "Nietzsche, Freud, Marx," published a decade later, Foucault suggests that psychoanalysis is a technique of interpretation.[10] As such, it is a kind of hermeneutics of suspicion, displacing the question of truth as being directed towards that which is manifest to that which is not seen, the invisible, that which lies below the surface. In his famous article "The Technologies of the Self,"[11] his use of the term technology refers to the ways in which a conception of the self has arisen through distinct corporeal and discursive practices. In another text, "Self writing,"[12] Foucault introduces his project as part of a series on "the arts of oneself" or on "the aesthetics of existence" or, even, on "the government of oneself and of others." The aim was to shift focus onto the first two centuries of the Roman Empire. This enquiry was then developed into the volumes comprising *The History of Sexuality*. Here, Foucault claims that the ancient "care of the self"—in which philosophy, and the truth-claims that belong to it—could be described as an ethos rather than a science, a way of life in which body and soul participate as equal terms. The ethos of the care of the self aims towards certain achievements; through these achievements one becomes true to oneself. Modern scientistic ideals, however, have replaced the notion that truth is something one achieves with the notion that truth is something one discovers. This, in turn, has contributed to the creation of forms of confessional practices of subjectivation. Psychoanalysis, according to Foucault, is one form of such subjectivation. In this sense, psychoanalysis is a form of interpretation that will also shape a certain relation of the self to itself, a relation based on confession. If we were to believe such a notion of psychoanalysis then, indeed, the question "who am I?" would have made perfect sense to psychoanalysis. As we have seen, however, this is not the case.

Foucault separates the technologies of the self, including what he calls the care of the self, from a hermeneutics of the self, which has served as a guide for philosophy in the West. The care of the self has, however, been forgotten and obscured. In his text on "The Hermeneutic of the Subject,"

---

[9] James Miller, *The Passion of Michel Foucault* (London: Harper Collins, 1993), 62.
[10] Foucault, *Aesthetics: Essential Works of Foucault 1954–1984*, Vol. 2 (London: Penguin Books, 2000), 269.
[11] In Foucault, *Ethics: Essential Works of Foucault 1954–1984*, Vol. 1 (London: Penguin Books, 2000).
[12] In Foucault, *Essential Works* Vol. 1.

Foucault recognizes a form of existential impulse or conversion through which the gaze of the self is developed towards a new aim, that of shaping the self through a set of social practices. These were called *askesis*. From the moment that these practices were put in place, the quest for rational discourse or theoretical forms of truth had started. The aim in late antiquity—the period in which these discourses begun to develop— is not, however, to discover the truth (as in the case with the hermeneutics of the self), but rather to link the subject to a truth, one that is learned, memorized and progressively put into practice. To progressively apprehend a relation to truth is to establish a relation to the external world through "a quasi subject that reigns supreme in us."[13] Perhaps such a "quasi subject" could be likened to a Freudian superego, or a Lacanian Other, to which the desire of the subject is directed.

In "The technologies of the self" Foucault names four kinds of technologies serving as the means by which man has come to understand himself: (1) the technologies of production, through which man produces or changes artifacts; (2) the technologies of sign systems, making it possible for us to create meanings through the use of signs and symbols; (3) the technologies of power, shaping and dominating the individual towards certain goals and (4) technologies of the self, through which the individual elaborates his intellect, body and feeling in order to achieve wisdom, happiness and other moral values. All these technologies were regarded by Foucault as interwoven. The technologies of power, through moral and social taboos, for instance, have come to influence the conception of the self. This in itself is a concept that is very much possible to look at in conjunction with the Lacanian notion of the subject as a product of the law. While investigating the tradition of the care of the self, Foucault became increasingly uninterested in the idea of discourse as regulative for the individual but was instead engaged in the individual's relation to himself. The technologies of the self have a particular place in the system of technologies, representing a position through which the individual is able to formulate a productive conception of the self:

> Perhaps I've insisted too much on the technology of domination and power. I am more and more interested in the interaction between oneself and others, and in the technologies of individual domination, in the

---

[13] Foucault, *Essential Works*, Vol. 1, 102.

mode of action that an individual exercises upon himself by means of the technologies of the self.[14]

Foucault defines the self as a striving towards knowledge. But such striving has not always looked the same. To Aristotle, there is a natural link between sensation, pleasure, knowing and truth. The self, in Aristotle, is a unity in which the discovery of truth becomes enjoyable; knowledge becomes something you should strive for since it enriches the self. The ethical injunction to care for the self—and for how the self is constituted—leads a development of the care of the self during a long time. The practices that are developed in late antiquity are specific in that they are aiming towards a freedom of the flesh: ascetic techniques are developed in order to win a freedom from bodily needs. The care of the self has, however, been overshadowed by the injunction of "knowing" oneself. In this way, an important insight is lost; there are different ways of creating knowledge, and there are different forms of self.[15] To a certain extent, however, this insight was rediscovered by Nietzsche. In Nietzsche, the self is something set in motion on a field where the will to power determines everything: here, the self is an aid to a deceitful will striving for advantage and survival. The question of knowledge is redefined as a question of technologies—technologies that are created to define truth, to hide truth and to create truth. These technologies are submitted to the will to power.

Foucault's point is that the notion of the self has, historically, developed through experiences allowing thereby the self to change. It is this insight that must be rediscovered. When new technologies have developed, the "care of the self" has also developed. For instance, the art of writing has been tied to introspection, through the writings of Marcus Aurelius. He shows how a new freedom of the self may develop through the nuances of introspection: in this, the very experience of the self is developed. Writing allows for a whole new field of experiences and possibilities to emerge. The care of the self does not just apply to the soul, but also to the body. A tradition has developed from Marcus Aurelius to Nietzsche: detailed descriptions of food and the body's reaction to the intake of food have become an important part of the male, philosophical tradition of diary writing. In this tradition, the technique of diary writing is described as making congruent the "gaze of the other and the gaze which one aims at oneself."[16] In the text,

[14] Foucault, *Essential Works*, Vol. 1, 225.
[15] Ibid, 228
[16] Ibid, 221.

the technology of the self is thus described as a form of *aisthesis* through which the pleasure of the body is made a tool for self-enquiry. Moreover, it is a process of exteriorization, through which the written sign makes one appear to the gaze of the other as well as the gaze of the self.[17]

Moral *paideia* was part of the *askesis* of the free man. In classical Greek thought, "the ascetics that enabled one to make oneself into an ethical subject was an integral part—down to its very form—of the practice of a virtuous life, which was also the life of a free man in the full, positive and political sense of the word."[18] Freedom, to the Greeks, entailed a certain form of relation of the individual to himself. Freedom, here, is the same as mastery of the self. To Socrates, the ontological recognition of the self to the self emerges through desire: through the need to know oneself in relation to desire and the need to subdue these desires. The relation of the self to truth is described not as a discovery but as a form of strife. The concept of truth was used as a moral measure of moderation and was not a question of the decipherment of one's inner being. This question is then displaced through the ontological writings of Plato. With Diotima, Foucault argues, we move from the question of what kind of object one is supposed to love, into the question of the true nature of love. "For Plato, it is not exclusion of the body that characterizes true love in a fundamental way; it is rather that, beyond appearances of the object, love is a relation to truth."[19] Socrates does feel erotic love, but the aim of his desire lies elsewhere than in the fulfillment of bodily needs. The real virility of Socrates lies in the fact that he knows the true object of that desire. Socrates is thus a double master: not only does he have dominion over his own body, he is also a master of truth.[20]

In his notion of the technologies of the self, Foucault argues that truth is something you may learn to achieve rather than something you may discover. This offers a shift in perspective in relation to the hermeneutics of the self. The subject is a product formatted outside of the norms of society, but it can take shape in relation to these norms in forms that also define its freedom. To Foucault, the question of technology is therefore closely linked to governmentality. In his text "On the Government of the Living" Foucault asks: "How is it that in Western Christian Culture the government of men demands, on the part of those who are led, not only acts of obedience and

[17] Ibid, 216.
[18] Foucault, *The History of Sexuality, Vol. 3: The Care of the Self*, trans. Robert Hurley (London: Vintage Books, 1990), 77.
[19] Foucault, *The Care of the Self*, 239.
[20] Ibid, 243

submission but also acts of truth, which have the peculiar requirement not just that the subject tell the truth but that he tell the truth about himself, his faults, his desires, the state of the soul, and so on?"[21] In the text, Foucault analyzes the practice of confession integral to the Christian heritage—the very practice of confession that, in *The History of Sexuality,* is considered to be developed in psychoanalysis. As we have shown, however, the question of the subject in psychoanalysis has less to do with confession than with transference: the analyst will not listen so much to what is being said, or being confessed, but rather will ask: "who is talking?" It is a question that traces out a structure in which the subject is the product of a law. If we implicate the Who as a particular instance of a discourse—in which the subject is split, dislocated, and displaced—it will always be submitted to such a law, a law that in turn is productive of the desire that manifests itself in the phenomenon of transference. It must be the case, then, that if Foucault is talking about the subject as a product of governmentality, it is a product of submission, of the laws and the norms that produce the splits and displacements that follow every attempt to reformulate the conditions of the self. However, in Foucault, the historical evidence that speaks in favor of the self is overwhelming: the split of the self is a question of technology and of development, not of structural necessity.

As we have shown, both Lacan and Foucault work with the assumption that there is a limited amount of freedom to be had for the split subject, but there is at least some. Moreover, both Lacan and Foucault work with the assumption that the concept of technology can be used in understanding how the self can be restructured in relation to the laws and the norms that will help produce it. The question that remains to be answered is: is there a concept of transference present in Foucault?

In her book, *Giving an Account of Oneself,* Judith Butler argues that the presentation of the self is always engaged in transference. It is not just an aspect of analysis, but of human interaction. The I addresses a you, and both are affected in the process. Transference alters the question "who are you?" because the You can never be known. It is both the you of the analyst and something beyond, and so the question is returned to the self; there is no Other that can answer the question. The question of the Other is not intertwined with the formation of the self. It is impinged from the outside before any self can appear. The first problem of psychoanalysis, then, is not the opening towards alterity. It is, rather, building an I from too much other-

---

[21] Foucault, *Essential Works,* Vol. 1, 81.

ness.[22] It is from this perspective that we must consider Foucault's critique against psychoanalysis. The problem of psychoanalysis is not the unconscious or the unknown. It is, rather, that the talking cure dissolves any possibility of forming a self. To Foucault the question is not that of the desire of the Other, but: "what can I become?" It is in formulating a self, in recapturing the possibility of a subject that is not split, that freedom can be recuperated.

---

[22] Butler, *Giving an Account of Oneself*, 72.

Enunciation and Politics
A Parallel Reading of Democracy

Maurizio Lazzarato

> When revolutionary discourse takes on the form of a critique of
> existing society, it plays the role of a parrhesiastic discourse.
>
> Foucault

Jacques Rancière claims that political subjectivation "never interested
Foucault, at least not on the theoretical level. He was occupied with power."[1]
This is a too hasty and careless judgment, since subjectivation constitutes
the very culmination of Foucault's work. In fact, we are here confronted
with two radically heterogeneous conceptions of political subjectivation.
Contrary to Rancière, for whom ethics neutralizes politics, Foucault's
political subjectivation is indistinguishable from *ethopoiesis* (the formation
of *ethos*, the formation of the subject). According to Foucault, the necessity
of connecting the transformation of the world (institutions, laws) and the
transformation of oneself, of others, and of existence, constitutes the very
problem of politics as it is configured after May 68. These two conceptions
of subjectivation are the expressions of two quite heterogeneous projects to
grasp actuality. This also comes across in the diverging readings of the
institutions and modes of functioning of Greek democracy proposed by
Foucault and Rancière. The two approaches display remarkable differences
not only with respect to the conception of the political, but also in relation
to language and enunciation.

For Rancière, Greek democracy has once and for all demonstrated that
the unique principle of politics is equality, and that, in language, we find the
minimum of equality necessary for a comprehension of living beings, per-
mitting us thereby to verify the principle of political equality. Speech,

---

[1] "Biopolitique ou politique?" interview with Éric Alliez, *Multitudes*, No. 1 (March 2000):
88-94, citation at 90.

whether it is in the form of command or of a problem, presupposes mutual understanding in language, and political action should increase and realize this potential, no matter how small, which is contained in language.

In Foucault's reading of the very same democracy, equality constitutes a necessary but insufficient condition of politics. Enunciation (truth-telling, *parrhesia*) creates paradoxical relations in democracy, since truth-telling introduces difference (of enunciation) in equality (in language), which necessarily implies an "ethical differentiation." Political action takes place within the framework of those "paradoxical relations" that equality entertains to difference, the culmination of which is the production of new forms of subjectivation and singularity.

Truth-telling (parrhesia)

Foucault enters into the problem of democracy through truth-telling (*parrhesia*), i.e. by way of how someone in the assembly takes the floor, rises up, and assumes the risk of enunciating the truth concerning the city's affairs. As an analyst of democracy, Foucault picks up a classic theme from one of his masters, Nietzsche: the question of the value of truth, of the will to truth, or of "who" it is that wants the truth.

The relation between truth and the subject is no longer posed in terms of his earlier work on power, namely in asking through which practices and through what types of discourse power attempts to speak the truth about the mad, delinquent, or imprisoned subject, or how power constitutes the "speaking, working, living subject" as an object of knowledge. From the end of the '70s onwards, the perspective shifted, and the formula now became: What type of discourse of truth is the subject "susceptible to and capable of applying to itself"?

The interrogation that traverses the lectures on Greek democracy is guided by a typically Nietzschean question, which in fact concerns our actuality: What does it mean to "tell the truth" after the death of God? Contrary to what Dostoyevsky had claimed, the problem is not whether everything is permitted, but if nothing is true, then how should we live? If the concern for truth consists in constantly turning it into a problem, then what type of life, which powers, which forms of knowledge and discursive practices, could support it?

The capitalist response to this question is the constitution of a "life market," where everyone can purchase the existence that suits him or her. It is no longer the philosophical schools, as in ancient Greece, nor Christianity, nor the revolutionary projects, as in the eighteenth and nineteenth centuries, that propose modes of existence and modes of subjectivation, but companies, the media, the culture industry, the institutions of the welfare state, and the unemployment insurance agency.

In contemporary capitalism, the governing of inequalities is strictly connected to the production and governing of modes of subjectivation. The contemporary "police" works both through the division and distribution of roles, and by the assigning of functions and the injunction to lead certain forms of life: each revenue, each assignment, each salary refers to an "ethos," i.e. to a way of acting and speaking, and it prescribes and implies certain conducts. Neoliberalism is at the same time the establishing of a hierarchy based on money, merit, heritage, and a veritable "market of lives," where companies and the state, which replace the master or the confessor, prescribe a conduct (how to eat, where to live, how to dress, how to love, how to speak, etc.).

Contemporary capitalism—its companies and institutions—prescribe a care of the self and a work on the self that is both physical and psychological, a "good life" and an aesthetic of existence that appear to delineate the new frontiers of capitalist subjection and evaluation, and which signal an unprecedented impoverishment of subjectivity.

Foucault's last courses are an indispensable tool for interrogating this. His analysis first of all requires that we should not isolate the political act as such, as Rancière does, because, Foucault claims, one then risks missing the specificity of capitalist power and the way in which it mobilizes politics and ethics, a non-egalitarian division of society, and a production of models of existence and discursive practices. Foucault invites us to connect the analysis of forms of subjectivation, the analysis of discursive practices, and the analysis of the techniques and procedures by which one undertakes to conduct the conduct of others. To summarize: subject, power, and knowledge must be thought together in their irreducibility and necessary interrelation.

In gradually shifting from the mode of political subjectivation toward the sphere of personal ethics and the constitution of the moral subject, *parrhesia* gives us the possibility to think the complex relations of these "three distinct elements, which cannot be reduced to each other [...] but whose interrelations are mutually constitutive."[2]

---

[2] Michel Foucault, *Le courage de la vérité: Le gouvernement de soi et des autres II, Cours au Collège de France 1984*, ed. Frédéric Gros (Paris: Gallimard/Seuil, 2009), 10.

Parrhesia, politeia, isegoria, dynasteia

In his two last courses, Foucault demonstrated that *parrhesia* (truth-telling), *politeia* (the constitution guaranteeing that all men endowed with citizenship are equal), and *isegoria* (the statutory right of anyone to speak, irrespective of social status, birthright, wealth, or knowledge) establish paradoxical interrelations. For *parrhesia* to exist, for truth-telling to be exercised, there must be both *politeia* and *isegoria*, which affirms that anyone can publicly take the floor and have his say in the affairs of the city. But neither *politeia* nor *isegoria* indicates who in fact will speak, who in fact will make a claim to truth. Anyone has the right to speak, but it is not the egalitarian distribution of the right to speak that makes someone really speak.

The effective exercise of *parrhesia* depends neither on citizenship nor on a legal or social status. *Politeia, isegoria*, and the equality that they declare, only constitute necessary—though insufficient—conditions for the act of speaking in public. What in fact makes someone speak is *dynasteia*: power, force, the exercise and real actualization of the power to speak, which mobilize singular relations between the speaker and himself, as well as between him and his addressees. The *dynasteia* that is expressed in enunciation is a force of ethical differentiation, since it means to take up a position in relation to oneself, to others, and to the world.

In taking sides and pitting equals against each other, and in bringing polemics and litigation into the community, *parrhesia* is a risky and indeterminate action. It introduces conflict, agonism, and dispute into public space, and may even lead to hostility, hatred, and war.

Truth-telling, the pretension to truth enunciated in an assembly (one can think of assemblies in social movements and contemporary politics, since Greek democracy, unlike modern democracy, is not representative), presupposes a force, power, or action over oneself (to have the courage to risk telling the truth), and an action over others, in order to persuade them, guide and direct their conduct. It is in this sense that Foucault speaks of an ethical differentiation, or a process of singularization, unleashed by parrhesiastic speech. *Parrhesia* implies that, by taking up a position, political subjects constitute themselves as ethical subjects, capable of putting themselves at risk, of posing a challenge, and of pitting equals against one another, i.e., as subjects capable of governing themselves and of governing others in a situation of conflict. In the act of political speech, in speaking in public, there is a power of self-positioning and auto-affection, where subjectivity affects itself—as Deleuze says in relation to Foucault's idea of subjectivation.

Parrhesia restructures and redefines the field of possible action both in relation to the self and to others. It modifies the situation; it opens for a new dynamic precisely in introducing something new. Even though the structure of *parrhesia* implies a status, it is a dynamic and agonistic structure that transgresses the egalitarian framework of right, law, and the constitution.

The new relations expressed by truth-telling are neither contained in nor foreseen by the constitution, law, or equality, and yet it is through them (and them alone) that a political action becomes possible and can actually be carried out.

Truth-telling thus depends on two heterogeneous regimes: right (*politeia* and *isegoria*) and *dynasteia* (power and force), and this is why the relation between true speech (discourse) and democracy is "difficult and problematic." In introducing a *de facto* difference into equality, in expressing the power of auto-affection and auto-affirmation, a double paradox is created. First, "there can only be a true discourse by way of democracy, although true discourse introduces into democracy something entirely different, irreducible to its egalitarian structure,"[3] i.e. ethical differentiation. Secondly, "the possibility that true discourse will die out and be reduced to silence"[4] is inscribed into equality, since dispute, conflict, agonism, and hostility threaten democracy and its equality. This is what in fact has happened in our western societies, where there is no longer place for *parrhesia*. Democratic consensus is the neutralization of *parrhesia*, of truth-telling, and of the subjectivation and action that flow from it.

Enunciation and pragmatics

The difference between the positions of Rancière and Foucault emerges even more clearly if we take a closer look at the relation between language and enunciation, politics and political subjectivation. For Rancière, the fact that those "who have no part" (*demos* or proletariat) begin to speak does not mean that they become conscious or that they express what properly belongs to them (their interests or a belonging to a social group). What it refers to is rather the equality of *logos*. The inequality of domination presupposes the equality of speaking beings, because if the order of the master is to be executed by the subordinate, the master and the subordinate must

---

[3] Michel Foucault, *Le gouvernement de soi et des autres: Cours au collège de France, 1982–1983*, ed. Frédéric Gros (Paris: Gallimard/Seuil, 2008), 167f.
[4] Ibid, 168.

be able to understand each other and share a common language. The fact of speech, even in those relations of power that are profoundly asymmetric (as in the discourse of Menenius Agrippa on the Aventine hill, which wants to give legitimacy to hierarchical differences in society), presupposes understanding in language, a "community whose law is equality."[5]

For an action to be political, one must first suppose a declaration of equality that functions as a rule and ground for the argumentation and the demonstration in a litigation between the rule (of equality) and the case (inequality of the police). Given that equality has been declared somewhere, its power must be realized. "Being inscribed somewhere, it must be expanded and reinforced." Egalitarian politics finds a source of legitimacy and a set of arguments in the logic and structure of language. Politics consists in the creation of a "scene where the equality and inequality of the parties in the conflict, considered as speaking beings, are put into play."[6]

For Rancière, there is indeed a logic of language, although it is negated by the duality of logos, "speech and counting of speech." Speech is at the same time the place of a community (speech that expresses problems) and of a division (speech that gives orders). Against this duality, political enunciation must argue and demonstrate that "there is one common language," and establish that the ancient *demos* just like the modern proletariat are beings that, simply by the fact of speaking and arguing, are beings endowed with reason and speech, and in this are equal to those that command: "The dispute does not bear on the contents of language [...] but on the consideration of speaking beings as such."[7] If Rancière plays with universals and discursive rationality ("the first requirement of universality is the universal belonging of speaking beings to the community of language"[8]), while also distinguishing himself from them, then Foucault describes subjectivation as an immanent process of both the rupture and the constitution of the subject.

For Foucault, *parrhesia*—to use a formulation from Félix Guattari— "steps out of language," but also out of pragmatics as it is understood in analytical philosophy. There is no discursive or logical rationality, because enunciation is not connected to rules of language or pragmatics, but to the

[5] Jacques Ranciere, *Aux bords du politique* (Paris: La Fabrique, 1998), 102. "The egalitarian logic implied in the act of speech and the inegalitarian logic inherent in the social bond" (115).
[6] Jacques Ranciere, *La Mésentente* (Paris: Galilée, 1996), 80.
[7] Ibid, 71.
[8] Ibid, 86.

risk of taking up a position, to an "existential" and political auto-affirmation. There is no logic of language, but an aesthetic of enunciation, in the sense that enunciation does not verify what is already there (equality), but opens for something new that is given for the first time in the act of the one who speaks.

Parrhesia is a form of enunciation very different from the one that pragmatics describes on the basis of performatives. Performatives are formulas, linguistic "rituals" that presuppose a more or less institutionalized status of the speaker, and in which the effect that the enunciation should produce is already given institutionally ("The meeting is open," enunciated by the one who is authorized to do it, is only an institutional "repetition" whose effects are known in advance). *Parrhesia*, on the other hand, does not presuppose any status; it is the enunciation of "anyone." Unlike performatives it "opens for an indeterminate risk," it provides "possibilities, a field of dangers, or at least a non-determined eventuality."[9]

The irruption of *parrhesia* determines a fracture, something that suddenly breaks open a situation, and "makes possible a certain number of effects" that are not known in advance. The effects of enunciation are not only always singular, but first of all affect and engage the enunciating subject.

The reconfiguration of the sensible first of all concerns the speaker. In the parrhesiastic statement [*enoncé*], the speaking subject forms a double pact with itself: it commits itself both to the statement and to the content of the statement, or both to what it has said and to the fact that it has said it. There is a retroactive action of the enunciation on the subject's mode of being: "In producing the event of the statement, the subject modifies or affirms itself, in any case determines and renders precise which mode of being it has as speaking."[10]

Parrhesia manifests the courage and the taking up of a position of the one who states a truth, who says what he thinks, but it also manifests the courage and the taking up of a position of the "interlocutor, who accepts to receive the painful truth he hears as true."[11] Who speaks the truth, and says what he thinks, "as it were signs the truth that he enunciates, he commits himself to this truth, he is obliged to it and by it."[12] But he also takes a risk "concerning the relation he has to the one he is addressing."[13] If the professor possesses a

---

[9] Michel Foucault, *Le gouvernement de soi et le gouvernement des autres*, 61.
[10] Ibid, 66.
[11] Foucault, *Le courage de la vérité*, 14
[12] Ibid.
[13] Ibid, 15.

"knowledge that is *techne*," and does not risk anything by speaking, the parrhesiast takes the risk not only of polemics, but also of "hostility, war, hatred, and death." He takes the risk of pitting equals against themselves.

Between the speaker and what he enunciates, between the one who tells the truth and the one who accepts to receive the word, an affective and subjective bond is established—a "belief," which, as William James points out, is a "disposition to act."[14] The self-relation, the relation to others, and the belief that unites them, can be contained neither in equality nor in right.

Crisis of parrhesia

In the crisis of Greek democracy Rancière perceives simply a desire of the aristocrats to reestablish their birthright, status, and wealth, whereas Foucault, without disregarding this aspect, sees the crisis as coalescing around the relation between politics and ethics, equality and differentiation.

The enemies of democracy put their finger on a problem that the proponents of equality as the sole principle of politics (Rancière and Badiou) do not see, and which constitutes thereby one of the stumbling blocks of nineteenth and twentieth century communism, without this having led to any useful responses.

As the enemies of equality claim, if everyone can have their say in the affairs of the city, there will be as many constitutions and governments as there are individuals. If everyone can take the floor, then the fools, the drunkards, will be authorized to state their opinions about public affairs in the same way as the best and those that are competent. In democracy, conflict and dispute, agonism and conflict among equals that all pretend to tell the truth, degenerate into seduction by orators who flatter the crowd in the assemblies. If the right to speak is handed out without control, "anyone can say anything." How can we then distinguish the good from the bad orator? How can we produce an ethical differentiation? The truth, enemies of democracy always claim, cannot be spoken in a political field defined by the "indifference between speaking subjects": "Democracy cannot make room for an ethical differentiation of subjects that speak, deliberate, and are capable of deciding."[15]

---

[14] See William James, *La volonté de croire* (Paris: Seuil, 2005) [*The Will to Believe, and Other Essays in Popular Philosophy*, 1897].
[15] Foucault, *Le courage de la vérité*, 46.

These arguments immediately call to mind the neoliberal critique of "socialist" wage equality, and, more generally, equality thought in terms of social rights: equality is an obstacle to freedom and "ethical differentiation," and it drowns subjectivity in the indifference of subjects of right.

In the same way as Guattari, Foucault cautions us that we cannot oppose neoliberal liberty, which in reality is an expression of a political will to re-establish hierarchies, inequalities, and privileges, solely by an "egalitarian politics." For this would be to disregard those criticisms leveled against egalitarian socialism by different political movements already before the liberals.

Foucault does not only denounce the enemies of democracy, but, drawing on the Cynics, he reverses the aristocratic criticism on its own terrain: ethical differentiation, constitution of the subject and its becoming.

After the crisis of *parrhesia* there emerges a "truth-telling" that no longer exposes itself to the risk of politics. Truth-telling in its origin derives from the sphere of personal ethics and the constitution of the moral subject, although it confronts us with the alternative between the "metaphysics of the soul" and the "aesthetics of life," between knowledge of the soul and a purification that gives access to another world, to other practices and techniques that might serve to test, experiment, and transform the self, life, and world as they exist here and now. This is the constitution of the self, no longer as "soul," but as "bios," as a form of life. This alternative is already contained in Plato's texts, but it is the cynics who render it explicit and, by politicizing it, turn it against the enemies of democracy. The opposition between the Cynics and Platonism can be summarized as follows: the first articulate "another life" and "another world," another subjectivity and other institutions in this world, whereas for the latter, the issue is rather "the other world" and "the other life" in a connection that will be prolonged in Christianity.

The Cynics reverse the traditional theme of the "true life," into which truth-telling had migrated and installed itself. The "true life" in the Greek tradition is a life that escapes perturbation, changes, corruption, degradation, and which maintains itself without modification in the identity of its being. The Cynics reverse this "true life" by claiming and practicing "another life," "whose alterity must lead to a transformation of the world. Another life for another world."[16] They reverse the theme of the "sovereign life, calm in itself and beneficial for others," into an "activist life, a life of

[16] Ibid, 264.

163

combat and struggle against and for oneself, against and for the others," [17] a combat "in this world against the world."[18]

In connecting politics and ethics (and truth) in an indissoluble way, the Cynics transcend the "crisis" of *parrhesia*, and the impotence of democracy and equality to bring about ethical differentiation. They dramatize and reconfigure the relation to the self politically, by wresting it away from the good life, and from the sovereign life of ancient thought.

Two models for political action

These two readings of Greek democracy are informed by two very different models for "revolutionary" action.

For Rancière, politics is a compensation for a wrong done to equality, through the method of demonstration, argumentation, and interlocution. Through political action, those that "have no part" must demonstrate that they speak and not just emit noise. They must also demonstrate that they do not speak another language or a minor language, but express themselves in, and master, the language of their masters. Finally, they must demonstrate by argumentation and interlocution that they are beings endowed with reason and speech.

The model for revolutionary action based on demonstration, argumentation, and interlocution aims at an inclusion, a "recognition" that, no matter how litigious, comes very close to a dialectical recognition. Politics calls forth the division into parts, where "we" and "they" are opposed as well as counted, where two worlds are divided while still recognizing that they belong to the same community. "The non-counted, in displaying the distribution [*partage*] by stealing the equality of the others, can make themselves be counted."[19]

If we were to find something that resembles Rancière's model, we should not look to democratic politics, but to the social democracy that was formed in the wake of the New Deal in the postwar period. This is the kind of social democracy that we still find in the French system of co-management of Social Security, the "dialectical model" of class struggle where the recognition and litigation between "us" and "them" constitutes the motor of development in capitalism and in democracy itself.

---

[17] Ibid, 261.
[18] Ibid, 262.
[19] Rancière, *La Mésentente*, 159.

That which Rancière defends in the social democracy of the welfare state is a public sphere of interlocution where the workers (the trade unions in their reformist version) are included as political subjects, and where work is no longer a private, but a public matter:

> One claims in bad faith that the institutions of welfare and solidarity that are born out of the workers' democratic struggle, and that are administered and co-administered by representatives of the contributors, are the products of a proliferating and paternalist state. In struggling against this mythic State, one attacks precisely those institutions of non-state solidarity within which other competences and capacities to take care of the common and of the future than those of governmental elites were formed.[20]

The major problem with Rancière's position (and more generally, that of the Left) is the difficulty of criticizing and surpassing the model that undoubtedly has broadened democracy in the twentieth century, but today is one of the major obstacles to the emergence of new political objects and subjects, since it is in principle incapable of including other political subjects than the state, the trade unions and the employers.

The political model that emerges from Foucault's analysis of Greek democracy is entirely different. Why does he refer to a philosophical school like the Cynics, a "marginal," "minoritarian," and "popular" school whose doctrines are relatively unstructured, in order to interrogate political subjectivation? What Foucault seems to suggest is the following: we have moved away from the dialectical and totalizing politics of the "demos."

For Rancière, "What has no part—the poor in antiquity, the third estate or the modern proletariat—can in fact have no other part than nothing or the whole."[21] However, it is difficult to see how the Cynics, like the political

---

[20] Jacques Rancière, *La haine de la démocratie* (Paris: La Fabrique, 2005), 91. I bought this book on the day of its release (September 2005) when I came back from an action undertaken by the Coordination des Intermittents et Précaires ["Association of Day-Laborers]. It was an explosive event, and they had occupied the room where one of these reunions in the Cultural Ministry was held, bringing together the state, the trade unions, and the employers, thus denying the status of political subject to everyone but themselves. Leafing through the book that same evening, I was struck when reading this passage. That liberals attack the welfare state is no reason for us to confine ourselves to a defensive attitude and silence the critique that different political movements directed against it in the '70s (it produces dependence and exercises power over the body), a critique which is continued today (it produces inequality, social and political exclusion, it controls the lives of individuals etc.).

[21] Rancière, *La Mésentente*, 27.

movements after '68 (the women's movement or that of the unemployed) would be able to claim, "we are the people," or, we are at the same time "the part and the whole." In Foucault's model, the problem is not to count those that have no part, or to demonstrate that they speak the same language as their masters, but it is to undertake a "transvaluation" of all values, which also, and first of all, concerns those that have no part and their mode of subjectivation. In this transvaluation, equality is connected to difference, and political equality to ethical differentiation. Through the Cynics we also once more encounter Nietzsche—these cynics that have entered the history of philosophy in the guise of "counterfeiters," as those that have altered the "value" of money.

The motto of the Cynics, "change the value of money," refers both to the alteration of money (*Nomisma*) and of the law (*Nomos*). The Cynics did not demand recognition, they did not want to be counted or included. They criticized the institutions and modes of life of their contemporaries, by experimentation with and testing themselves against others, and against the world.

The problem of how to constitute the self as an ethico-political subject also involves particular truth-games:

> No longer the truth-game of apprenticeship, of the acquisition of true propositions and knowledge, as in Platonism, but the truth-game that bears on oneself, on what one is capable of doing, on the level of dependence one attains, on the progress to be made [...] These truth games do not derive from the *mathemata*, they are not things that are taught and learned, they are exercises performed on the self – exercise, testing of oneself, a combat in this world.[22]

The political truth-games practiced in the constitution of another life and another world are thus no longer those of recognition, demonstration, and argumentative logic, but those of a politics of experimentation that binds together both the rights and formation of an *ethos*. The opposition between Plato and the Cynics is to some extent reminiscent of the one between Rancière and Foucault.

---

[22] Foucault, *Le courage de la vérité*, 210.

Logos and existence, theater and performance

For Rancière, politics exists only through the constitution of a "theater," a stage on which the actors perform the artifice of political interlocution by means of a double logic of discursivity and argumentation that is at once reasonable (since it postulates equality) and unreasonable (since this equality exists nowhere).

For there to be politics, a stage of "speech and reason" must be constructed by the interpretation and theatrical dramatization of the gap between the logic of police and of equality. This is a normative conception of politics. No action that understands public space other than as interlocution by way of speech and reason is political. Rancière, therefore, does not consider the actions of the suburbans in 2005, who did not respect this model of mobilization, as political.

> The issue is not to integrate people who, for the most part, are French, but to make them be treated as equals. [...] It is to know if they are counted as political subjects, endowed with a speech in common. [...] This movement of revolt has apparently not yet found a political form, as I understand this, i.e. the form of a scene of interlocution that recognizes the enemy as part of the same community as itself.[23]

In reality, contemporary movements do not neglect to actualize the political logic described by Rancière, i.e. to construct a scene of speech and reason so as to lay claim to equality by way of demonstration, argumentation, and interlocution. But even though they fight for recognition as new political subjects, they do not make this form of acting into the only one that could be defined as political. And even more importantly, these struggles take place in a framework, which is no longer that of the dialectics and the totalization of the *demos* that is at once part and totality, both "nothing" and "everything."

On the contrary, in order to impose themselves as political subjects, they are obliged to break away from the politics of the "people" and the "working class," such as it is incarnated in the political and social democratic regimes of our societies.

These political movements play and juggle with these different modes of political action, but according to a logic that is not limited to the staging of

[23] Jacques Rancière, "La haine de la démocratie – Chroniques des temps consensuels," *Libération*, Dec 15, 2005.

"equality and its absence." Equality is the necessary—though insufficient—condition for the process of differentiation, where the "rights for everyone" are the social basis for a subjectivation that sets in motion the production of another life, another world.

The "little savages" of French suburbia, as they were called by a socialist minister, in some respects resemble the "barbarian" cynics who, instead of engaging in the orderly and dialectical games of recognition, prefer to leave the theater stage and invent other artifices that have little to do with theater.

Rather than a theater stage, the Cynics call to mind contemporary performance art, where public exposure (in the double sense of manifesting and taking a risk) does not necessarily occur through language, speech, and signifying semiotics, or through the theatrical dramaturgy of casting, inter-locution, and dialog.

How does the process of subjectivation, which opens a path to "another life" and "another world," occur? Not simply through language and reason. The Cynics are not only "speaking beings," but also bodies that no doubt enunciate something, even though this enunciation does not pass through signifying chains. To satisfy one's needs (to eat, to shit) and desires (to masturbate, to make love) in public, to provoke, scandalize, to enforce thoughts and emotions, are different "performative" techniques that engage a multiplicity of semiotics.

The stick, the beggar's bag, poverty, wandering, begging, the sandals, the naked feet etc., by which the Cynic's mode of life is expressed, are modalities of non-verbal enunciation. The gesture, the act, the example, behavior, and physical presence are practices and semiotics of expression that address others without passing though the relay of speech. In Cynical "performances" language does not solely have a denotative and represent-tative function, but above all an "existential function." It affirms an *ethos* and a politics, and takes part in the construction of existential territories, to speak with Guattari.

In the Greek tradition, there are two ways to attain virtue: the long and easy way that passes through the logos, i.e. discourse and its scholarly apprenticeship, and the short but difficult way of the Cynics, which is "somehow mute." The short way, or shortcut, without discourse, is that of exercise and of putting oneself to the test.

Cynical life is not public only through language, through speech. Rather it exposes itself in its material, everyday reality. It is a materially and physically public life that immediately reconfigures the divisions consti-tutive of Greek society: on the one hand the public space of the *polis*, on the other hand the private administering of the house. The point is not to oppose "logos" with "existence," but to take up a place in the gap between them, in order to interrogate modes of life and institutions. For the Cynics, there can be no true life that would not be another life, at the same time of a "form of existence, self-manifestation, a plastic truth, as well as the task of demonstrating convincing, and persuading through discourse."[24]

In Rancière, just as in most of contemporary critical theories (Virno, Butler, Agamben, Michon) there is a logocentric prejudice. Despite Rancière's criticism of Aristotle, we are always dependent on, and within the framework of, the formulas of Greek philosophy: man as the only animal endowed with language, and a political animal because he has language. When the Cynics attack the "distribution" that the logos establishes between man and animal, they attack the foundations of Greek and Western philosophy and culture:

> In ancient thought, animality was the point of absolute differentiation for the human being. It was by distinguishing himself from animality that the human being affirmed and manifested his humanity. Animality was always a point of repulsion in this constitution of man as a human being endowed with reason.[25]

The Cynics do not dramatize the gap between equality and police, but instead place emphasis upon the practices of "true life" and its institutions, by displaying a shameful and scandalous life, a life manifesting itself as a challenge and an exercise of a practice of animality.

The distribution of the sensible, or Division and production

In Rancière, political subjectivation implies an *ethos* and truth-games. It requires a mode of constitution of the subject through language and reason, which practices the truth-games of "demonstration," "argumentation," and interlocution. Even in Rancière, politics cannot be defined as a specific activity, since it is articulated in relation to ethics (constitution of a subject of reason and speech) and to truth (discursive practices that demonstrate and argue). It is difficult to see how it could be otherwise.

[24] Foucault, *Le courage de la vérité*, 288.
[25] Ibid, 244.

But if it is impossible to make politics into a mode of autonomous action, it is also impossible to separate politics from what Foucault calls the microphysics of power.

In Rancière, the "distribution of the sensible," which organizes the distribution of parts (the division of classes that separates the bourgeois—who have speech—and the proletarians—who only express themselves through noises—as well as the mode of subjectivation ("they / we"), does not seem to allow for this type of relations. The distribution of the sensible is a division of functions and roles, modes of perception and expression, and it is produced in a double way where micropolitical relations play a fundamental role. The division of society into "classes" (or parts) is produced by an assemblage of discursive practices (knowledge), by techniques of governing conducts (power), and by modes of subjection (subject). But this "dualist" distribution is not only the result of the transversal action of these three apparatuses (knowledge, power, subject), it is itself traversed by relations of micropower that make it possible and operational. The relations between men and women, father and children in the family, teacher and student in the school, doctor and patient in the health system etc., which are developed within what Guattari called "public facilities" of subjection, are transversal and constitutive for the division into parts. There is a "molecular" distribution of the sensible, a microphysics of power that also traverses those that have no part (and divides them according to different lines than those of the great dialectical distribution of us and them). It is impossible to understand contemporary capitalism without interrogating the relation between the molar (the great dualist oppositions between capital and labor, rich and poor, those who command and those who obey, those who are entitled to govern and those who are not) and the microphysical (the power relations that even find support, pass through, and take shape within those that have no part).

"Bios," "existence," and "life" are not vitalist concepts to which one could oppose the concepts of the political division of the *demos*. They are rather domains of microphysical power, in which we find struggles, disputes, subjections, and subjectivations.

The reflection on how the Cynics understood *bios*, existence, and life, can provide us with the armature to resist the powers of contemporary capitalism, for which the production of subjectivity is one of the most important features. In a certain way, we are obliged to use the Foucauldian methodology because, in contemporary capitalism, it is impossible to separate "ethics" from "economy" and "politics."

Foucault tells us that the gradual shift of *parrhesia* from the "political" domain to individual ethics "is nevertheless useful for the city. In inciting you to care for yourself, it makes you useful for the whole city. If I protect my life, it is precisely in the interest of the city."[26]

The techniques for the formation of *bios* (the techniques for governing oneself and others), which were integrated and reconfigured by the pastoral power of the Christian church, have constantly become more important in the actions of the welfare state.

In capitalism, the "great chain of care and solicitude," the "care for life" of which Foucault speaks in relation to ancient Greece, is assumed by the state, at the same as it sends packing the population for the slaughter in war. To take care of oneself, to perform a work on oneself and one's own life, means to care for the ways of doing and speaking that are necessary for us to occupy the places to which we are assigned in the social division of labor. To take care of oneself is an injunction to become a subject responsible for that function to which power has assigned us. These techniques of constitution and control over conducts and forms of life are primarily experienced by the "poor" today (the unemployed, those with minimal income, poor workers etc.) The question posed by the concepts of *bios*, existence, and life is not that of vitalism, but how to politicize these relations of micropower by a transversal subjectivation. If not everything is political, as Rancière affirms—since "in this case politics would be nowhere"—then everything "can be politicized," Foucault adds.

At the level of a theoretical definition of politics, Rancière seems to neglect that which he analyzes from a historical point of view: the work on the self, the formation of an *ethos* that he elsewhere describes so magnificently in the case of the workers of the nineteenth century.

The formation of *ethos*, *bios*, and existence practiced by the cynics is not a version of "moral discourse." It does not constitute a new pedagogy or the vehicle of a moral code. The formation of an *ethos* is at the same time a "focal point of experience" and a "matrix of experience," where different forms of possible knowledge, the "normative matrices for individual behavior," and the "modes of virtual existence for possible subjects" are articulated in relation to each other.

In Rancière, politics is not primarily an experience, but above all a question of form. "What makes an action political, is not its object or the place in which it occurs, but solely its form, which inscribes the verification

---

[26] Ibid, 83.

of equality in the institution of a dispute, a community that only exists by way of division."[27]

The interrogation of these "focal points of experience" and the experiments that take place in the wake of the cynics, are transmitted throughout the history of the Occident, and are taken up in a renewed form by the revolutionaries at the end of the nineteenth and the beginning of the twentieth century, and by the artists from the same period.

Organization and activism

The Foucauldian subjectivation is not only an argument about equality and inequality, a demonstration of the wrong done to equality, but a veritable immanent creation that takes place in the gap between equality and inequality, displacing the question of the political by opening the indeterminate space and time of ethical differentiation and of the formation of a collective self.

If politics is indistinguishable from the formation of an "ethical" subject, then the question of organization becomes central, even if in different fashion than in the communist model. The reconfiguration of the sensible is a process that must become the object of an "activist" work that Guattari, prolonging Foucault's intuition, defines as an "analytical" political work.

For Guattari, the GIP (*Groupe d'information sur les prisons*) can be considered as an analytical and activist collective assemblage, where the object of "activism" is re-doubled: it belongs to the domain of intervention, but also to the side of those that intervene. The task is to permanently work with the statements produced, but also with the techniques, procedures, and modalities of the collective assemblage of enunciation (of the organization).

Inversely, Rancière has no "interest in the forms of organization of political collectives." He only considers the "alterations produced by the acts of political subjectivation," i.e., he only perceives the act of subjectivation in its rare emergence, whose temporal duration comes close to instantaneity.

He refuses to consider "the forms of consistence of groups that produce it,"[28] although May 1968 was precisely about interrogating their rules of constitution and functioning, their modality of expression and of demo-

---

[27] Rancière, *La Mésentente*, 55.
[28] Jacques Rancière, *La méthode de l'égalité: La philosophie déplacée* (Paris: Editions Horlieu, 2006), 514.

cracy, since political action and intervention are inseparable from the act of constituting a subject.

If the paradoxical relations of equality and inequality can be inscribed neither in a constitution nor in laws, if they can be neither learnt nor taught, but only experimented with, then the question of the forms of acting together becomes fundamental.

What happens when someone takes the floor, what happens afterwards? How does this act of differentiation affect not only the speaker, but also the one who accepts him, i.e., how is a community formed that is bound together through enunciation and artifice, and which would not be closed around its self-identification, but remain open to ethical differentiation?

What must be experimented with, and invented, in a war machine that mobilizes a being-together and a being-against, is that which Foucault claims to be the specificity of philosophical discourse, and which, due to the emptying out of the dialectical model of the *demos,* has become the condition of politics today.

One should never pose "the question of *ethos* without at the same time interrogating the truth and the mode of access to truth that could form this *ethos,* and the political structure within which this *ethos* could affirm its singularity and difference [...] never pose the question of *aletheia* without at the same time reopening the question of *politeia* and *ethos.* The same thing applies to the *politeia,* and to *ethos.*"[29]

In Rancière, only democracy as the apparatus of both division and community, can reconfigure the distribution of the sensible, whereas Foucault is much more reserved and shows less enthusiasm for this model of political action, since he perceives its limits. Political subjectivation, while resting on equality, also goes beyond it. The political question would then be how to invent and practice equality under these new conditions?

Translation: Sven-Olov Wallenstein

---

[29] Foucault, *Le courage de la vérité,* 63.

Rancière as Foucauldian? On the Distribution of the Sensible and New Forms of Subjectivities.

Adeena Mey

In his own words, Rancière's *method* resembles Foucault's. But, even if only in passing, Rancière has also touched on some of the divergences existing between his own work and Foucault's. These aspects can be found in *La Mésentente*, along with two interviews—the first of which was conducted by Eric Alliez and the second with one of his translators, Gabriel Rockhill.[1] Among the major points sketched in these texts—and on which this paper will be based—is Rancière's brief but frank criticism of the notion of biopolitics. The aim of the present paper is not to produce a systematic commentary on the similarities and differences that can be said to exist between these two thinkers, but rather to discuss Rancière's criticism on the basis of an empirical case, namely contemporary claims made around autism as a form of subjectivity. The scope of the paper is thus not exegetical. Rather, what it shall seek to discuss is the operativity of Rancière's critical remarks in fields studying subjectivities such as autism. Yet, as the criticism leveled at Foucault by Rancière is a corollary of a discrepancy present between their respective methods, and, moreover, as such a difference will be addressed through the lens of a central category from the latter's theoretical repertoire—namely, the "distribution of the sensible"—it will be necessary to start with a brief account of its most fundamental points.

---

[1] Jacques Rancière, *La Mésentente: Politique et Philosophie* (Paris: Galilée, 1995); "Bio-politique ou Politique? Entretien recueilli par Eric Alliez," <http://multitudes.samizdat.net/Biopolitique-ou-politique>, last accessed May 12, 2009; "Interview for the English Edition (with Gabriel Rockhill)," in Jacques Rancière, *The Politics of Aesthetics: The Distribution of the Sensible* (London and New York: Continuum, 2004), 47-66.

Rancière: The distribution of the sensible

On a few occasions, Rancière has explicated some similarities, as well as divergences, characterizing both his own and Foucault's methods. One such site for disagreement is what both thinkers mean by "politics." In *La Mésentente*, Rancière formulates a critique of biopolitics, extending Foucault's understanding of the "police," from which a critique of this concept can be formulated. Indeed, *La Mésentente* is an attempt to think the specificity of the political, which requires, Rancière says, a strict distinction from the "police." What he calls "the distribution of the sensible" is the principle governing our sensible order, creating shared understandings of what is visible and sayable. Also, as suggested by the original French *partage du sensible*, this principle of governing both divides but also creates common parts within the sensible, and by extension, modes of participation within this order:

> I call the distribution of the sensible the system of self-evident facts of sense perception that simultaneously discloses the existence of something in common and the delimitations that define the respective parts and positions within it.[2]

Within this ordering of the sensible, Rancière distinguishes two logics, and which can only be sketched here. First, basing his argument on Foucault's essay "Omnes et singulatim: Towards a critique of political reason," Rancière describes a logic that sees harmony in a given ordering of bodies, of their visibility or invisibility, and of the modes of saying and doing. This order is that of the "police." Rancière agrees with Foucault's analysis on the point that the "police" as a form of government extends beyond what he calls "the lower police" (the police of policemen and their sticks) and is thus part of "a social apparatus in which the medical, assistance and culture were entangled" and "bound to become a form of counselor, manager as well as an agent of the public order."[3] Yet, at the same time, Rancière distinguishes the order of the "police" from a second logic, which consists in the "suspension" of this order deemed harmonious. It is from the suspension of this given ordering of bodies and the way they participate in it as appearing and being heard that "politics" emerges. For Rancière the latter results from the suspension of the harmony of a sensible partition:

---

[2] Jacques Rancière, *The Politics of Aesthetics*, 12.
[3] *La Mésentente*, 51. All translations from this text are mine.

"Politics" should be used exclusively to characterize a clear-defined activity, which is also antagonistic to the former—the police. This activity, is one that disrupts sensible configurations in which shares, parts or their absence are defined in regards to the presupposition that there is, by definition, no share: the share of the share-less ones.[4]

The conditions for the appearance of the political are organized around a specific terminology that denotes dissensus and what Rancière calls "the wrong," in contrast to the consensual order of the police. Unfortunately, due to the limited scope of this paper, I will not be able to expand on this, but we have to retain that, in their most generic acceptance, the notions that Rancière develops denote the emergence of conflicting positions within the sensible, which question the very terms of those positions, producing thereby a conflict over the very definitions of those modes of saying and doing. In other words, politics doesn't emerge on a plane of actuality ordered by the police, but from the *presupposition* that another logic exists, that of equality, and that certain singular events confirm its existence. Thus politics does not exist *per se* but only from the encounter of the police and equality. In order to think this encounter one has to abandon certain concepts, the first of which is power. Indeed, for Rancière:

> The concept of power leads to the conclusion that if everything is police-related (*policier*), everything is political. Thus the following negative consequence: if everything is political, nothing is. If, as Foucault did, it is important to show that the order of the police extends beyond its institutions and specialised techniques, it is equally important to affirm that no thing is in itself political by the mere fact that power relations are exercised.[5]

Biopolitics is not politics

From this distinction between the police and politics, Rancière is also able to extend his critique to the concept of biopolitics and its contemporary vicissitudes. He formulates two problems. First, according to Rancière, an understanding of biopolitics and biopower, such as the one developed by Giorgio Agamben, has brought these notions into a domain alien to

---

[4] Ibid, 52-53. Hereafter, with respect to the terms being defined, I shall drop the use of quotation marks for "police" and "politics," except when quoted as such.
[5] Ibid, 55-56.

Foucault, namely that of the "modes of living" (*modes du vivre*), which is based on Agamben's wider attempt to bring Foucault closer to the concerns of both Arendt and Heidegger.[6]

It is principally on the second point of the critique, however, that I wish to focus on here. It concerns the modes of governing bodies, subjectivities as well as forms of interventions on health and disease—or to borrow the sociologist Nikolas Rose's expression, "life itself."[7] Even though Rancière does not explicitly engage with the latter, such a take on Foucault's work is well evoked by Rancière when he writes, for example, that this conception of biopolitics consists in granting a "positive content" to the notion, based as it is on an ontology of life, and remaining theoretically proximal to Deleuze's vitalism. For Rancière, this confounds *political subjectivation* with *processes of individual and collective individuation*.[8] Thus, within Rancière's theoretical frame, the many fields that have come under the influence of Foucault's work, and have been analyzed through the lens of biopolitics, are not therefore political but belong instead to the order of the police. As put by the French philosopher Mathieu Potte-Bonneville, it is as if "Foucault's perspective could just fit within Rancière's."[9] Hence, the idea that processes of individuation might be confounded with political subjectivation directly points to methods that are often a resort for fields studying contemporary subjectivities—methods that could then be "contained" within a Rancièrian frame. Trends in the history, sociology, and philosophy of psychological disciplines, could be domesticated too. Indeed, the latter examples are much indebted to Foucault's work and biopolitics often acts as a transversal notion. In this regard, the case of the "autism rights movements" and so-called "neuro-minorities" can serve as a discussion ground upon which both biopolitics and its Rancierian critique find embodiment.[10] Here I would like to resort to the example of (bio-)political claims and identity

---

[6] "Biopolitique ou Politique?," 2.
[7] Nikolas Rose, "The Politics of Life Itself," *Theory, Culture & Society*, Vol. 18(6) 2001: 1-30; idem, *The Politics of Life Itself: Biomedicine, Power and Subjectivity in the Twenty-First Century* (Princeton and Oxford: Princeton University Press, 2007).
[8] "Biopolitique ou Politique?," 2.
[9] Matthieu Potte-Bonneville, "Versions du politique: Jacques Rancière, Michel Foucault," *La philosophie déplacée: Autour de Jacques Rancière* (Paris: Horlieu, 2006), 180. To my knowledge, this text is among the very few works that systematically engage with Rancière's and Foucault's conceptions of politics.
[10] "Autism (rights) movement," "neuro-diversity" or "neuro-minority" are often used indiscriminately by the members of these communities despite their different origins. I prefer to resort to "neuro-minority" for its emphasis on the minority dimension, which translates well both my empirical and theoretical interests.

claims made around autism as a form of contemporary subjectivity, in order to show how such phenomena call for a critical assessment of our critical and analytical tools and how Rancière's method can help formulate these problems.

The autism rights movement and neuro-minorities

Since its first description by Leo Kanner in 1943 until the early 1980's, autism remained a very rare condition and was considered one of the most severe forms of child psychosis. But from the 1980s onwards, there has been a proliferation of cases, with the diagnosis of autism now extending to include non-severe forms, such as the conditions known as "high-functioning autism" and Asperger syndrome, i.e. autism without mental retardation, and generically defined today as impairment in socialization. Moreover, from a clearly defined psychiatric entity, autism has been reorganized on the model of a continuum, namely the "autistic spectrum disorders" (ASD)[11]. These changes have notably taken place, on account of the rise and mobilization of associations of parents of autistic individuals around research seeking a neurobiological or genetic basis of autism against psychodynamic ones—mostly psychoanalytical. Indeed, the latter have been attacked for the negative conceptions of mothers they have advanced, as is summed up, for example, by the expression "refrigerator-mothers." Underpinning such a label is the idea that mothers are principally responsible for the autism of their children.[12] The aim of this alliance between parents and scientists has been the search for a cure for autism, and for better care to those who suffer from the disorder.

Since the 1990s, and in parallel to this type of parental activism calling for a right to health, another type of claim has started to appear. These new claims originate from autistic individuals themselves situated on the high-functioning end of the autistic spectrum. Known as the "autism rights" or the "anti-cure" movement, these self-proclaimed neuro-minorities struggle for the recognition of autism neither as an illness nor as a handicap, but as *a*

---

[11] American Psychiatric Association *DSM IV. Diagnostic and Statistical Manual of mental disorders*, (Washington DC: APA, 1994). At the time of writing, debates on the redefinition of several categories (including ASD) of the DSM are taking place, which will lead to the fifth revision of the manual. This is likely to produce changes in the identity/political ecology of autism.

[12] The most iconic being Bruno Bettelheim's *The Empty Fortress*.

different way of being.[13] For these autistic individuals, no cure is needed for autism, as for other neuro-minorities (such as people with ADHD), as it is a cognitive and cerebral variation that simply exists within humanity. Similarly to racial and sexual diversity, neuro-diversity should be accepted, thus curing it would equal curing gay or colored people. The conditions of possibility for such claims can be understood within a larger frame, namely the emergence of bio-subjectivities and identities, which take the brain as their reference. Moreover, notions such as Paul Rabinow's "Biosociality," Nikolas Rose's "Biological Citizenship," "Neurochemical Self," and "Bio-subjectivity" have been formulated, mostly by anthropologists or socio-logists. All have enabled a greater emphasis and more sophisticated theorization of the ever-growing entanglement between the spheres of life— as defined by the life sciences—and life as experience, on its social, political, and juridical levels.[14] In the case of autism, while controversies still exist, genetics and neurobiology have offered semantic references to thematize common features that give rise to autistic subjectivities. While defined as impairment in socialization, skills associated with autism have made possible the advancing of claims for its acceptance as a different way of being. Indeed, if autistic individuals suffer from a lack of social intelligence necessary to socialize within society at large, they are thought to have a higher cognitive intelligence.[15] Moreover, regarding biosocial factors, autistic traits are widely found within populations of mathematicians and scientists, as suggested by the colloquialism "geek syndrome," and also with respect to ongoing studies as to whether Einstein, Newton but also Warhol had, or had not, Asperger syndrome.[16] Of course, not all individuals on the spectrum are savants, but the mere reference to brains capable of superior "sequential thinking" allows for claims to skip over any mention of dis-

[13] See for instance Michelle Dawson, *We Are Not Your Community: In Response to Autism Society Canada's Open Letter,* <http://www.sentex.net/~nexus23/naa_asol.html>, last accessed September 1, 2009.
[14] See the two texts by Nikolas Rose (as in note 7), and Bernard Andrieu, "La fin de la biopolitique chez Michel Foucault," *Le Portique,* 13-14 (2004), <http://leportique.revues.org/index627.html>, last accessed September 1 2009; Francisco Ortega "The Cerebral Subject and the Challenge of Neurodiversity," *BioSocieties,* Vol. 4 (2009): 425-445.
[15] Nicholas Putnam, *Kids Called Nerds: Challenges and Hope For Children With Mild Pervasive Developmental Disorders,* year not indicated, <http://www.aspergersyndrome.org/Articles/Kids-Called-Nerds--Challenge-and-Hope-For-Children.aspx>, last accessed August 29, 2009.
[16] See for instance Muhammad Arshad, & Michael Fitzgerald, "Did Michelangelo (1475–1564) have High-Functioning Autism?," *Journal of Medical Biography,* Vol. 12 (2004): 115-120; Steve Silberman, "The Geek Syndrome," *Wired* 9/12 (2001).

ability. This strict focus on talents and skills, alongside the statement that autism is simply an "alternative cognitive style," has almost made it a desirable way of being. As a contemporary figure of subjectivity, autism has shifted into the terrain of singularity and uniqueness. Furthermore, claims are made for the acceptance of autism as a mere difference, in the name of a popular understanding of humanism. Indeed, as one can read on the website of TAAP's (The Autism Acceptance Project, a self-advocacy project), accepting autism is about "tapping into human potential and dignity," while "the joy of autism" could "redefine ability and quality of life."[17]

The biopolitics or politics of neuro-minorities?

The case of autism and the claims concerning the acceptance of difference raise the following questions. First, claims surrounding identity are effectively locatable at the intersection of the biological, genetic, social and psychological and allow, through the reference to the brain, the contestation of "normality" as bio-socially normative. Indeed, for members of neuro-minorities, we are living in a neuro-typical world, neurotypicality being ironically defined by the autistic community as "a neurobiological disorder characterized by preoccupation with social concerns, delusions of superiority, and obsession with conformity."[18] Moreover, neuro-minorities' references to other identity-based social movements, as well as its neo anti-psychiatry accents and humanistic claims, add to its emancipatory dimension. Yet, since its actual potential to suspend a harmonious order requires better scrutiny, one should remain critical towards the latter. Indeed, the emphasis on the uniqueness and singularity of the autistic condition, paired with its cerebral ontological substrate, actually makes neuro-minority people representatives of one of the anthropological figures of contemporary individualism enabled by neurosciences. Indeed, difference is here conceived in neurobiological terms while the conception of subjectivity is paired with the plasticity of the brain, a conception, which is for Rose, "bound with more general norms of enterprising and self-actualizing."[19] Autistic subjectivity thus echoes late capitalist's imperative to become a flexible subject. But paradoxically, it also echoes formulations addressing the identity politics of minorities, as articulated by Foucault himself and for

[17] http://www.taaproject.com/
[18] http://isnt.autistics.org/
[19] Nikolas Rose, "The Politics of Life Itself," 18.

181

whom the affirmation of a minority required the "creation of new forms of lives and cultures."[20] Both situations share the same coordinates and precisely echo what Rancière describes when he states that Foucault's method is too bound by its "schema of historical necessity" and thus rendering certain things unthinkable. Indeed, as Rancière writes:

> I would say that my approach is a bit similar to that of Foucault's. It retains the principle from the Kantian transcendental that replaces the dogmatism of truth with the search for conditions of possibility. At the same time, these conditions are not conditions for thought in general, but rather conditions immanent in a particular system of thought, a particular system of expression. I differ from Foucault insofar as his archaeology seems to me to follow a schema of historical necessity according to which, beyond a certain chasm, something is no longer thinkable, can no longer be formulated.[21]

This thing—in the case of autism but also certainly for most processes of subjectivation—concerns, on the one hand, the political potential at work in strategies of self-definition. Such a potential should go beyond mere resistance. On the other hand it concerns the very theoretical apparatus and the intellectual tools we use to describe and analyze them. This double-bind both relates to empirical and theoretical/methodological aspects. One must effectively ask, first, if neuro-minorities suspend our sensible order and if the claims around autistic subjectivity can give rise to political subjectivation or if they simply reiterate positions within a consensual order. The second point, and an important corollary, relates to methodology. In an interview conducted by Rancière in 1977, Foucault, speaking of his famous metaphor of theory as a "toolbox" stated that it meant, notably, producing thoughts on *given situations*. He added that such research was "necessarily historical regarding some of its dimensions."[22] Today, following Foucault's steps, Rabinow and Rose have attempted to safeguard the concept of biopower from Agamben on the one side and Hardt and Negri on the other. In their enterprise, they stated that biopower—including biopolitics in the same schema—should "designate a plane of actuality."[23]

[20] Michel Foucault, "Sex, Power and the Politics of Identity," in *Ethics: Essential Works of Foucault 1954–1984,* Vol. 1 (London: Penguin Books, 1997), 166.
[21] Jacques Rancière, "Interview for the English Edition," 50.
[22] Michel Foucault, "Pouvoir et stratégies: Entretien avec J. Rancière," *Dits et Ecrits II* (Paris: Gallimard, 2001), 427.
[23] Paul Rabinow & Nikolas Rose, "Biopower Today," *BioSocieties* 1 (2006): 197.

Doubtless, biopower, biopolitics and other "bio-" conceptual tools forged in a Foucauldian mould are very accurate in describing and analyzing situations of "governmentality" in which action upon life and practices of the self are at stake. But to analyze and to formulate political subjectivation, as Rancière understands it, or any other form of minor subjectivity that produces conflicting positions, we must allow ourselves a shift of focus and question thereby the relevance of the historical and the empirical. Such a shift might be seen as the site where boundaries between the actual and the virtual are negotiated. As suggested by the example of autism, describing but also producing knowledge on a plane of actuality renders processes of individuation barely discernable from what seems to be political subjectivation, even more so when the former is biopolitical. If we agree that political subjectivation takes place in a heterological mode, then perhaps the latter should become the site of our very theoretical tools too.

Formulated in a different context and drawing on Deleuze, the following thought from Mariam Fraser nevertheless sheds light on our problem. Effectively, as the sociologist has shown, the empirical is not, in itself, a guarantor of *relevance*. Rather, relevancy is gained when a problem serves as a "lure" for a virtual problem. She thus calls for the possibility to submit research problems to virtual rather than social and historical structures.[24] The virtual is precisely, for Rancière, what allows us to think the unthinkable, which cannot take place within the order of the police. The virtual requires one to think and to do "as if" (*comme si*). Indeed, Rancière writes, "the political is the production of a theatrical and artificial sphere."[25] Moreover, historicism can only relegate the possible to its temporal dimension, only foreseeing other modes of existing in near-future occurrences. The virtual, on the contrary, is superimposed on the given world. Yet, this does not proscribe Foucault, nor does it attempt to play on a straightforward opposition, Foucault versus Rancière. On the contrary, one can follow Foucault who—as Potte-Bonneville reminds us—saw his own work as "philosophical fragments put to work in historical fields of problems," and—as much as his perspective fits within Rancière's[26]— those fragments could, virtually and through an act of superimposition, contain questions brought by Rancière's philosophy too.

---

[24] Mariam Fraser, "Experiencing Sociology," *European Journal of Social Theory*, Vol. 12(1) (2009): 63-81.
[25] Rancière, *The Politics of Aesthetics*, 4.
[26] Potte-Bonneville, "Versions du politique," 179-80.

Roundtable

Sven-Olov Wallenstein

Biopolitics is, as we know, a theme that appears at a certain point in Foucault's work, and then disappears, or is, rather, absorbed into other concerns—for instance the problem of governmentality, of conduct and counter-conduct, and later into the idea of subjectivation and the technologies of the self, to mention two of the most visible ones.

While this may seem to give the topic of biopolitics a lesser importance, it is also true that it constitutes something of a caesura in Foucault's work, just after the first volume of the *History of Sexuality* (where it appears for the first time in the published work, as a kind of addendum). The emphasis that we find in *Discipline and Punish* on processes of discipline as pervasive in modernity does not disappear altogether, although it is fundamentally modified with the introduction of the apparatuses of security, which have a certain situated freedom as their correlative, and together make up something that at least in the 1977–78 lectures on *Security, Territory, Population* can be called a kind of biopolitical complex.

It is in this context that Foucault makes the suggestion that liberalism is the fundamental form of governmentality within which biopolitics unfolds, first by way of an analysis of its development in the eighteenth century, but then, in following the 1978–79 lectures on *The Birth of Biopolitics*, more surprisingly also during long in-depth discussions of modern neoliberalism.

As a way to open for the general discussion, I would like to pose three general questions that I think have been present throughout the talks and the discussions, but that need to be stated even more clearly, perhaps even bluntly:

1. The first question has to do with what could be called *historical specificity*. When Foucault says that we are still within the kind of problem that was initiated in the eighteenth century, how should we understand

this? Do biopolitics and liberalism form some kind of *longue durée* of political modernity, in the same sense that Foucault seemed to be proposing earlier with respect to discipline? How can we make room, within the conceptual structure that Foucault proposes, for that which undeniably also separates us from the Enlightenment and its understanding of govern-mentality? He was often criticized for portraying discipline as a kind of all-encompassing structure that leaves us with no way out—and to some extent liberal biopolitics may seem to present us with an even more seamless narrative, which extends all the way up to the present. The discourse of discipline, however, always had as its flipside the idea of resistance, even that "resistance comes first," as Foucault said, whereas liberalism and bio-politics—precisely because of the emphasis that each places on freedom—seem to make the idea of an outside even more difficult to grasp, unless we would opt for unfreedom or some sheer irrationality.

2. The second question bears on Foucault's more precise understanding of *freedom*. It is rather clear that Foucault is not posing the classical meta-physical question of free will, nor is he following a Kantian path that would locate freedom in the relation between a faculty of desire and a moral law, but as a situated concept that only exists in correlation to other complexes of knowledge of power, as in the case of the apparatuses of security. But later he will also speak of freedom as connected to subjectivation and technologies—even an "aesthetic"—of the self, although mostly with reference to ancient Greek and Roman texts. What, then, would be the technologies of the self that exist in the modern neoliberal world? What is the role of agency in a neoliberal world where the freedom of choice seems more like an enforced freedom? Once more, the question of resistance imposes itself.

3. The third question, finally, relates to the problem of *ontology*. Are there implicit ontological assumptions in Foucault's work, and if so, do they shift through his various phases? In the analysis of discipline, it seems to me that there is an underlying idea of the body as a source of resistance—the body as an assemblage of affects in the sense of Deleuze, that will always overflow the disciplinary framework, so that we "do not yet know what a body is capable of," as Deleuze used to say with reference to Spinoza. But in the theory of biopolitics this idea of an underlying and as it were onto-logically resistant multiplicity seems to, if not disappear, then at least fades into the background. Life is now that which a certain governmental rationality discovers as a source for its operations, it is the *ratio essendi* of politics as knowledge—from the Physiocrats onwards—and not something

that engenders an immediate, or even prior, resistance. Can we say that Foucault understands the question of "life" as a properly ontological question, or does he simply historicize this concept as yet another invention within power-knowledge, which has no priority as such?

So, these were the three questions I would like to put, as simply as possible. Perhaps we should just start with the first one.

Johanna Oksala

This question was posed to me already yesterday, and I didn't manage to give a very good answer then. But I have now had some time to reflect, and maybe I understand the question better. You are asking: is there an outside to this liberal governmental regime of truth, or this liberal governmentality, and if so, what would this outside be? Or in more practical terms, how does one resist neoliberal governmentality? In my paper I argued that neo-liberalism was a much deeper and more complex phenomenon than a mere economic doctrine, and that this entailed a fundamental rethinking of the tools of critical thought as well as of political resistance. But it is my contention that the neoliberal production of regimes of truth is never complete, nor is their operation as internally consistent as neoliberalism's own representations would lead us to think. We must question their hege-mony, as well as the political neutrality of economic knowledge, and analyze the way in which economic truths produce political effects. We must also advocate the seemingly crazy argument that the maximal material well-being of the population is not necessarily the undisputed aim of good government. At the moment I am very interested in these political move-ments of "degrowth" or "post-growth" that aim for global well-being with-out relying on economic growth to make it happen. In other words, we should question neoliberalism's exclusive claim to rationality and regain and reinvigorate alternative political values, such as justice and equality, with which to assess the ways we are governed. While we have to accept that practical forms of resistance against neoliberalism have to consider the efficaciousness of their strategies and even apply strictly economic, cost-benefit analysis to some of their actions, economic rationality should, and does not form the only framework for assessing politics. So, there is obviously an outside. I also think that Marcia Sá Cavalcante Schuback's question yesterday was to the point: we tend to look at politics from a very westernized perspective. There are alternative governmentalities elsewhere in the world. Perhaps the most paradigmatic counterweight at present can be found in the Islamic world. Whether we like it or not, Islamic theocratic government represents a very different kind of political rationality, a different kind of governmentality. According to Islamic law, you cannot pay or charge interest, for example. Such a principle would completely devastate our governmentality since our political system relies so heavily on banking and global financial markets. However, I am obviously not suggesting that Islam would provide a solution to our problems. But at least one way forward in imagining political alternatives could be a dialogue with the rest of the world—with these alternative systems of thinking about politics, society, and culture.

Maurizio Lazzarato

I would like to bring up the question of freedom. There are many limits to Foucault's analysis and we have to shed some light on them. When it comes to liberalism, there is never any question of money, which is very astonishing. Liberalism, or more precisely neoliberalism, is really a question of finances. There is no discussion of property in Foucault's analysis, which is another very important limitation. There is a discourse on freedom in liberalism that is always connected to a discourse on property. In the theories of liberalism one is free only to the extent that one has property. If liberal theory had in fact been implemented, there would for instance never be such thing as voting, or what we understand as voting, because in all theories of liberalism voting is connected to having property. It was the labor movement that made a system of voting possible, which was never an issue for liberalism. Foucault says many things that are very imprecise, and we can never take them literally or accept what he says at face value. Sometimes he makes an apology for liberalism, for instance in the *Birth of Biopolitics*. When, for example, he raises the theory of human capital he sees only the relation in which the worker becomes an entrepreneur of himself. But at the same time this is a production of workfare. I think it is precisely this ambiguity in Foucault's work that we have still to shed light on. A concept of freedom or liberty is always very ambiguous, because we don't know what it means to be free; we are never simply free, but always also caught up in relations of dependence. A sociologist like Gabriel Tarde proposes a theory that is in fact much more interesting. He says that one can always act or choose differently, but he doesn't say that we are free. I think we have to take the history suggested by Foucault in a kind of reverse manner, and I find much more interesting things in Deleuze and Guattari on the relation between liberalism and freedom and capitalism. Deleuze and Guattari claim that capitalism is always characterized by two antagonistic movements: one is the hyper-modernity of capitalism, its hyper-innovative character; on the other hand there are a series of neo-archaisms that emerge. These things go together. This is why, in the society of communication, spaces of freedom go together with George Bush! In Italy, to offer a further example, you have a kind of capitalist alliance between the media system and Berlusconi, and which goes together with xenophobic political organizations like the Lega Nord. In France you have the modernist discourses of Sarkozy that goes well together with The Ministry of Immigration and National Identity. So, there are these obvious limits to Foucault's discourses on liberalism and we have to understand them, otherwise we end up in danger of becoming François Ewald.

Julian Reid

I think, in response to Maurizio, it is wrong to think of the absence of an account of the role of property and finance in Foucault's theory of liberalism as a weakness as such. I think it was a deliberate decision on his part not to address liberalism through those well-trodden tropes, but to think about liberalism specifically as a regime of power as opposed to a regime of exploitation, profits and loss. There's a brilliant discussion between Deleuze and Foucault, where Foucault literally says that, you know, we know where the exploitation occurs, we know where the profit goes, but we don't know how to explain the powers through which these regimes of exploitation and profit sustain themselves. That's a very different and much more complex problem and way of approaching liberalism than the traditional Marxist one. I don't think it's true to say that liberalism only permits freedom in so far as we possess property. I think it's more complicated than that. Freedom is conceptualized *as* a property, as a biological property and a capacity of the human. Liberalism aspires to governance in so far as it can know and regulate the exercise of freedom as a property of the body, and as a capacity of the biohuman. So I want to say that we should avoid the risk of treating Foucault as either just another or "the new" Marx. Let's not reify these texts or expect them to deliver answers to, or complete descriptions of, questions about how liberalism is functioning today, how it works, what its basic principles are, even what its ontology is—if it indeed has an ontology? I mean, it seems obvious to me that we still live in a biopolitical world—we've never been so biopoliticized. When I look at the character of international relations and the role and functions of the discourses of life—especially in the governing of the world of peoples—it seems that biopolitics is becoming a more and more endemic kind of problem. But at the same time we have to work with Foucault's concepts, and I think the concept of biopolitics is an important one to work with. We have to leave behind or test the limits of Foucault's own analytic. And one of the ways in which we can do that is by posing the question "what is the bio today?" I don't mean that we should pose that question ontologically, but do as Foucault did, and examine empirically how life is being discursively constructed by liberal regimes of power. Because I think that the account of life at stake in contemporary neoliberalism is very different from the account of life that was at stake within classical liberalism, and even the neoliberalism of the seventies, eighties and nineties. One of the ways in which I see this occurring is in the context of the growth of so-called sustainable development policies, and the ways in which discourses of sustainability and environmental crisis are being invoked to govern people, and in governing them, interpellating them within markets, market-based systems of governance, property rights regimes, and neoliberal practices of self-subjectification. So in a sense, the question of the nature of the "bio" to be secured via liberal governance has shifted from the classical liberal concern with the life of human populations to the life of the biosphere. And we are increasingly governed in ways that are designed to deny fundamentally human capacities to transcend our environments and develop instrumental, even exploitative, relationships with nature in order to protect the biosphere while at the same being interpellated within markets and neoliberal systems of subjection. And these developments, I think, challenge many of our essential assumptions about liberalism and neoliberalism. I mean, it's not entirely clear to me that neoliberalism is anymore a regime concerned fundamentally with increasing economic productivity. It seems to me we're moving into a regime of neoliberalism which governs us and keeps us in our places by denying us the capacity to secure ourselves through the economic means of our choice, by denying us the capacities to increase our productive abilities and establish some kind of secure relationship with the worlds on which we depend—all of which, I would say, are fundamental attributes of what it is to be human. So in other words, to resist biopower and biopolitics today, I think we have to revisit the question *of* the human; maybe to attempt to rescue a more fundamentalist account or understanding of what it is to *be* human. And what it is indeed to have stuff like, for example, security. It would be wrong to treat security as a universal, even though Foucault himself does that at certain points in his texts. When he makes that absurd claim that freedom is nothing but a correlate of security, he's treating freedom and security in universal terms, in complete contradiction to his supposedly fundamental ontology of concepts as practices. I think we can rework concepts like the human, like security, revalorize them in ways that can be politically productive and used to speak back to regimes of neo-liberalism, which are fundamentally concerned with governing us by styling us as posthuman subjects.

Johanna Oksala
I think we have already moved over to the second question, the question of freedom, and how we can resist the freedom that is in fact forced upon us—what was the formulation?

Sven-Olov Wallenstein
Enforced freedom.

Johanna Oksala
Enforced freedom. If power produces the neoliberal subject as "free," what would it mean to resist this? To be unfree? Like you said, this makes no sense. I think that one obvious way to circumvent this problem is to see that the "freedom" that is forced upon us is a very reductive and restrictive understanding of freedom. Similar to Julian, I also hold that it is not simply connected to property. It's equated with a freedom of choice. Yesterday I was very much emphasizing the economic growth argument—the neo-liberal project has advanced because the economic growth argument is such a powerful political weapon—but of course it has also relied on the classical liberal ideal of freedom. Market mechanisms must be left alone, not only because then we'll all be rich, but because then we'll all be free—free markets guarantee that people have maximal choice in cheap products and services. This freedom—the freedom of choice—effectively masks the systemic aspects of power by relegating the subject's freedom to a choice between different options whilst denying one any real possibility of defining or shaping those options. So, I don't think the answer is to give up the political ideal of freedom. We can still strive to be free, but freedom must be understood to mean something else than the freedom to choose between cheap products that we don't need. We must rethink what freedom means, but we must hold on to it and not let the neoliberals hijack it.

I also want to comment on the point that has been made about life. Catherine suggested earlier, and Julian suggested now, that in order to really to think about biopolitics we need to understand what life means today. So we need biology, we need a more sophisticated understanding of life. Now, that's fine, but I want to be polemical, and since I am a philosopher (and not a natural scientist), I think that it is not life that we need to try to understand, but politics. So biopolitics—the problem of biopolitics—cannot be reduced to a biological question concerning our understanding of life, it has to concern our understanding of politics. In my view, the essential problem is that we have come to believe that politics is a science, that politics has become something like a science. I might suffer from philosopher's arrogance, but I think that it is crucial to see that politics cannot be a science. The problem of biopolitics is, in my view, exactly the idea that we could somehow move political problems to the realm of bioscience. Instead of turning to biosciences, we should try to rescue the philosophical question of what it is that makes human societies into political communities. This was Aristotle's famous question: why is a political community different from a colony of ants? I think it was interesting when Catherine said that the ghost of Aristotle always comes up when we talk about biopolitics. I think it *should* come up, because we must repeat his question.

Catherine Mills

I might respond to that if I can. The first thing to clarify for me is that I would never want to attribute the question "what is life" to Foucault. I think that would be a mistake, so that's certainly not my intention. I'm not even sure that I would actually take on that question as my own question, to be honest. I don't think the question is so much "what is life?"—but rather: What does the "bio" of biopolitics actually reference today? What would be the referent of that prefix? And there are a couple of reasons why I think it is important to answer that particular question. If we look at the ways in which the techniques of governance that manage life and extract value from it—and the kind of bio-evaluation that happens within biopolitics—if, that is, we only look at that conception of life, which is used within those techniques, then the risk is that one ends up with a kind of reductionist, objectivist conception of life that actually continues to contribute to those techniques. Now, what I think can actually be done in a different way is to look not just at biology to tell you what life is, but to look within biology for alternative conceptions and tools for thinking about what life *can be*. So, what could the referent to "bio" be, if we take it in a different and more affirmative way than just as a technique of management? So, this is to offer one reason. And in this sense, I think the question "what is life" is actually an impossible question to really take on and answer in the sense of giving an ontological response. I want to take the kind of epistemological paradox, which arises from our ineluctable being-in-language, very seriously (that is, that we can't know the world separately from what we say about it and consequently, what we know of the world is discursively "constituted"), but, from here, the important question for me is how one might respond to this paradox. You can make claims about the nature of the world, but of course we're always operating within language and there's a disjunction between language and the world. Do we then respond by trying to withdraw altogether from ontology, saying that we can't make ontological claims, or we can't say anything about the world? Or do we rather respond and say, well, we have to make these claims in some way, because they are actually politically important claims to make? This latter approach seems more compelling to me, though always with the caveat that the claims we may make are culturally and historically located claims, they are limited by and have to be understood within their own location (and where that location may not be exactly the same as the location of the speaker). So in that sense, I think, the point for me is not to answer the question "what is life?" but rather to answer the question: what does the "bio" reference? What political, social, or theoretical work is that prefix doing? And, furthermore, what other tools are there for thinking about the "bio" that might actually give us a more affirmative way of thinking politics than those mobilized within biopower? It may also be worth noting here that I am not suggesting some kind of valorization of "natural" life over and against "political" life. My point is not about a return to the "natural," as if that were even possible. The question of biology and referent of the "bio" is, then, not a question of the natural as opposed to the cultural, for example.

And as for Johanna's point about politics: another reason for looking at the question of the bio, is actually to try to identify what is in fact specific about biopolitics. I mean, I think you're right, I think the question of politics is extremely important and I would not suggest for one moment that we should turn to the idea of life as opposed to politics, or that we should reduce politics to a science, and so on. But it seems to me that if we use biopolitics without any understanding of what the "bio" actually means, then biopolitics simply collapses into politics and the concept itself becomes meaningless. There's no reason to use a concept of biopolitics unless the bio

FOUCAULT, BIOPOLITICS, AND GOVERNMENTALITY

actually references something that makes it distinct from other forms of politics. So in that sense we ought to try and make sense of the term, or stop using it. I personally think it's a valuable way of thinking about some of the things that happen today, so I think it's worth trying to make some sense of it.

Finally, let me make one last point in relation to the question of freedom, which in any case is not unrelated to the issue of the meaning of "bio." I agree with some of the points that have been made so far. Parenthetically, Nikolas Rose makes some nice points about the difficulty of enforced freedom in his book *Powers of Freedom*; there he develops the idea of being obliged to be free under neoliberalism. Given this obligation to be free, the question, it seems to me, is not "how do we resist freedom?" but what work does "freedom" do, why this freedom and not that, and how can the concept be mobilized differently? So the question is not "how do we resist freedom?" but "how do we rethink what freedom can be?," such that we can enact freedom in a different way. And it seems to me that there are actually resources even within the liberal tradition that might help in this regard. I'm thinking, for example, of Isaiah Berlin and the distinction he draws between negative and positive freedom. Of course, Berlin's distinction is confused and problematic in various ways, but it might nevertheless be useful. Negative freedom is typically seen as a lack of external restraint that relates both to property and choice, and positive freedom is a kind of autonomy or mastery. But positive freedom can be turned towards a kind of practice of the self, if you like, a kind of making of oneself. This is essentially what positive freedom would entail. Understood in this way, freedom has to be something other than just the supposed expansion of choice—which is often little more than a form of abandonment—and a correlative insecurity; this demands that conditions are in place to allow people the space and capacity for self-formation.

So I think there are ways of appropriating concepts, and doing so even from within those traditions that one might actually want to oppose; sometimes one can find tools for that opposition as well. This would be quite in line with Foucault's "rule" about the "tactical polyvalence of discourse." Finally, I'm fully cognizant of the fact that biology is very much part of biopolitics. But the point is to look at the tradition of biology and try to find resources there for thinking about life in a different way. Or, perhaps to look at the tradition of liberalism and try to find resources there for a different kind of freedom from the one enforced upon us.

Vikki Bell

I have to confess to a certain frustration now. It's strange for me that we come to these questions of freedom and the subject and resistance and so on—themes that have been so central to feminist critique (discussions that are so rich and have been going on for so long). Sorry, but I feel I just need to mention that before I say what I want to say. Because I think that a lot of these questions, including the question of whether you take up the questions of life or not, depend on what you think your work is for. So, for me, these questions we are addressing are very much relative to the idea of critique, as well as to the idea of genealogy. The politics of genealogy is not necessarily contained in the genealogy. So you wouldn't necessarily look for the answers—what should be done? how should we resist freedom?—in Foucault. And you wouldn't necessarily find the answers even if you undertook the most brilliant genealogy. But you might. So it's very important to me to think about the idea of critique and the idea of resistance separately. And this is why Foucault isn't a Marxist. The two would have to be dislodged somehow. So, the area that I have been writing about in relation to this question of life and vitalism relates more to the concept of performativity and the critique of performativity that is presently emanating from what we're calling the "new vitalism." This bears precisely on the question of freedom and the unfolding, as developed out of the Deleuzian reading of Foucault. And there I think, to try to sum that up, it relates to the question of the "bio," the question of "life." What is the critique of performativity? In some ways it's a playing out of the relationship between Deleuze and Foucault. Why do people start to critique, let's say, Judith Butler's work on performativity, just to give it a name. What I've argued is that the critique of performativity is a form of preformism, in the sense that the subject can only unfold what exists, if you like, discursively. And if it already exists discursively, then all the subject can do is unfold these possibilities that exist discursively. Then what you've removed is precisely the creativity. And to me that's why we have to talk about life in relation to politics. Because if you have a form of preformism in which individuals arise, then what you've done is in a sense to give up any political movement, because you've removed the movement from the virtual to the actual. And if you don't believe that there is anything there to be said about that process of creativity, then you are in that kind of preformist moment. And while that could be true, I think that Foucault's movement in the second and third volume of the *History of Sexuality* was precisely to allow that possibility, to allow for life to be creative. So I don't know if that's clearer or not, but that's my way of trying to relate the idea of life as creative to an idea of politics through the process of subjectification.

Thomas Lemke

First of all I would like to thank you, Sven-Olov, for stressing these three points. I really think you've captured the most important points and the most difficult questions to answer. Concerning the relation between freedom and agency I think that there is some kind of slippage of vocabulary, which is also present in Foucault's work. It is necessary to distinguish agency from freedom. Freedom of choice is a very limited and specific freedom—a liberal concept of freedom, which is characterized by consumer choices and by a certain spectrum of possibilities of action. At the same time, though, you have to choose, you are obliged to choose. However, we are not confronted with the paradox of an "enforced freedom" (I was tempted to say of "enduring" freedom). The task is rather to decipher this very specific kind of freedom, the very format of liberal freedom, in order to map its limits and costs. Also, I don't share Maurizio's, interpretation of Foucault's lectures on governmentality as some kind of apology of neoliberalism. I think that Foucault tried to analyze the inventiveness of liberalism and neoliberalism, and to contrast this inventiveness with socialism. He did this by addressing the question of why there's no socialist governmentality. He was fascinated with the fact that something real has been invented, a system of thought and a set of practices. To reconstruct this process also means to learn from it, and to be able to imagine a different governmentality. And this was what he was referring to—too optimistically, perhaps—when the socialists won the presidential election in 1981. The first interview he gave after the election was about the possibility of a new *logique du gauche*, a logic of the left. And I think this was really what he was trying to do: to learn from neoliberal inventiveness in order to imagine something different than neoliberalism.

Maurizio Lazzarato

I want to be a little bit provocative. One has to historicize neoliberalism, I think. There's a first phase, which ends at the time of the first Gulf war, in which you find the innovative and productive aspect of liberalism. And then there is a second phase, a phase of decline. Foucault couldn't grasp this because he was writing *The Birth of Biopolitics* in a preceding period. It's a fantastic book, a fabulous book to read, but still you have this shift where everything changes after the Gulf war. One has to introduce other things into this history. A very interesting case to discuss in relation to the idea of an administration of choice would be the medical reforms that Barack Obama is trying to infuse in the US. There is something like forty-five million people who have no access to medical care, others who have limited access to medical care because they have limited property. And then there are those who are against Obama's reform, saying this is a socialist reform that prevents us from making choices. This is because if they had the property, they'd have the money. What this example really shows, is that the problem of choice is the problem of money.

Thomas Lemke

I want to return, very briefly, to the first point you were making, about the question of resistance: how can you resist productivity? I think the most important point would be to ask: What do we mean by resistance, especially since liberal governmentality takes into account forms of resistance; resistance is, after all, not something exterior to liberalism, challenging it from the outside. It is part of the productivity and mobility of liberalism, there's a permanent process of response, adaptation, and reformulation. We should avoid the idea of a stable and fixed totality that remains unchanged, unchallenged—in fact it's permanently challenged. And we have to make visible the points of friction and the points of transformation that too often escape from the analysis. And doing this would mean to reinscribe conduct and counter-conduct into this very productivity of liberal governmentality. As for the "bio"—the bio of biopolitics, I think there are several ways to address the problem. Let me mention just two of them. One would be to ask how it comes to be that the biological—the reference to the body—is so important in contemporary forms of government? The other would be to further investigate and imagine the *bios*—something that would be a more comprehensive concept of life, and one not necessarily limited to a biological idea, and where, yes, the biological may have a role, but not the dominant and central role that it has today in politics.

Sven-Olov Wallenstein

I think it is time to have some comments from the audience. Please, the floor is open.

Marcus Doverud
I would like to ask a question that relates to the idea of neovitalism. You spoke of the solicitation to continuously individuate or create subjectivations as an imperative of freedom, but also an imperative of *jouissance*. If we cannot *not* continue to individuate, if we are solicited to perform ourselves anew, might this form an urgency or a fear? We can be driven by fear, but I would also like to think that there is another approach to being solicited to individuate. And this I guess is an attempt to further pursue your question: "is there an outside of neoliberalism?"

Vikki Bell
Should I respond to you? I think that you're right, and if I understand you correctly you're saying that perhaps we should remember the little moment that Foucault draws our attention to in the *History of Sexuality*: bodies and pleasures. Perhaps we are focusing too much on bodies and not enough on pleasures. And I think this is right, that it calls for sociological work, perhaps of the kind that Maurizio is doing. It would be boring just to look for resistance without understanding the complexity of... let's just call it emotion, the complexity of different emotions that are co-present in something we might want to call resistance. But you could well approach it in another way. In Argentina, for example, in some of the work that I was touching on yesterday, it's quite difficult to pose the question: "What is the pleasure of mourning?" And yet it is imbued in this work I was showing. Even the making of the posters, which are about something that is so upsetting for people, has nonetheless an aesthetic to them; and, yes, there's a pleasure in making them and putting them up. And I think it's a very complicated question, one that sociology doesn't handle very well. And if we reduce the empirical work we do to simply looking for resistance in the sense that we've been talking about, then I think you're right that we do miss something, and it's something we could relate to Deleuze and the idea of affect, and so on.

Jonna Bornemark
I would like to hear a bit more about the third question, which was posed initially, the one on ontology. Is there a relationship between genealogy and metaphysics, and would Foucault take the direction of a possible metaphysics, for instance in the sense to which Deleuze comes close?

Johanna Oksala

Thanks, Jonna, I've been waiting for this ontology question, since—well, you know, it's my favorite question, and I've written a lot about this. I promise that I won't go on forever, but I *could* go on forever. To begin with, I think it is ridiculous to claim that there is no ontology in Foucault, if by ontology we mean the tacit background beliefs about reality on which our thinking operates. In everyday life, as well as in philosophy, we constantly operate on the basis of assumptions that cannot be empirically proved or disproved. In this sense, everybody has an implicit ontology, and this is clearly the case for Foucault too. We can of course discuss what these implicit ontological background beliefs are, and because they're implicit, this is an endless debate. But I want to make a stronger claim here. I think that Foucault engages with ontology, not only in the sense that he too operates on the basis of implicit ontological background beliefs, but in the sense that he's actually making at least one really important and *explicit* ontological claim: power produces reality. All ontological orders are outcomes of different power relations, historical processes, and political struggles. Ontology, our tacit and taken for granted understanding of reality, is always an outcome of a political struggle: ontology is politics that has forgotten itself. Foucault's genius lies in making this process visible—in politicizing aspects of our present. His analyses have initiated new schemas of politicization: by uncovering new kinds of relations and mechanisms of power he has brought new questions and areas of experience such as insanity, delinquency, and sexuality into political debate. In exposing concepts, categories, and practices as sedimentations and expressions of power relations he attempted to reveal the exclusion, domination, and violent treatment of those at the losing end of the struggle for objectivity and truth: how their views have been branded as false and irrational and their behavior as abnormal and pathological. So, I think that if we argued that there was no ontology in Foucault, we would miss what was his most important contribution to political thought and activism. I know that contending that Foucault makes ontological claims upsets many Foucauldians—when I've given papers on Foucault's politicization of ontology some people always get upset and I must say that I am really puzzled about this. I can understand why people from a social science background might be reluctant to accept this claim, but I've never understood why philosophers would want to read Foucault and hold that he makes no ontological claims. If this is how they want to read Foucault, why don't these philosophers then do history or social science?

FOUCAULT, BIOPOLITICS, AND GOVERNMENTALITY

Fotis Theodoridis

I agree with you that there is an ontology in Foucault's work, but I think it is implicit, which is also my problem. And I could accept also the definition you provided here. But you just removed the problem, which is the ontology of politics.

Johanna Oksala

What do you mean by the ontology of politics?

Fotis Theodoridis

I mean, if ontology is what is produced by politics, or by relations of power, why do we have politics?

Johanna Oksala

That's an excellent question, and to answer it, we have to dig up at least some of the implicit ontological claims that Foucault is making. I think one of them is the irreducibility of power relations. There is one text in which Foucault says it explicitly: "I cannot imagine a society without power relations." Power relations are a kind of an ontological constant. And from this premise we get the other ontological claims, the productive nature of power and the fundamentally political nature of reality.

Authors

Maurizio Lazzarato is sociologist living and working in Paris. His recent publications include *La fabrique de l'homme endetté: Essai sur la condition néolibérale* (Paris: Editions Amsterdam, 2011), *Expérimentations politiques* (Paris: Editions Amsterdam, 2009), *Intermittents et précaires* (Paris: Editions Amsterdam 2008, with Antonella Corsani), and *Le gouvernement des inégalités: Critique de l'insécurité néolibérale* (Paris: Editions Amsterdam, 2008).

Thomas Lemke is Heisenberg Professor of Sociology with focus on Bio-technologies, Nature and Society at the Faculty of Social Sciences of the Goethe-University Frankfurt/Main in Germany. His research interests include social and political theory, biopolitics, social studies of genetic and reproductive technologies. Recent publications include *Der medizinische Blick in die Zukunft: Gesellschaftliche Implikationen prädiktiver Gentests* (with Regine Kollek, 2008), *Governmentality: Current Issues and Future Challenges* (ed. with Ulrich Bröckling and Susanne Krasmann, 2010), *Biopolitics: An Advanced Introduction* (2011), and *Foucault, Governmentality, and Critique* (2011).

Helena Mattsson is an Associate professor in Architecture and History and Theory of Architecture at the KTH School of Architecture, and Vice Dean of the School of Architecture and Built Environment, KTH (Royal Institute of Technology). Her doctoral thesis was published 2004, *Arkitektur och konsumtion: Reyner Banham och utbytbarhetens estetik* (Architecture and consumption: Reyner Banham and the aesthetic of expendability). She has written extensively on architecture, art and culture, and her publications include *Swedish Modernism: Architecture, Consumption and the Welfare State* (d. with Sven-Olov Wallenstein) 2010) and *1%* (2006). Mattsson is currently working on the project *The Architecture of Deregulations: Post-*

modernism and Politics in Swedish Architecture (together with Catharina Gabrielsson). Mattsson is an editor for the culture periodical *Site*.

Adeena Mey is a Swiss Science Foundation doctoral researcher at the University of Lausanne and a research fellow at Ecal, Lausanne University of Art and Design, Switzerland. His research interests include psychiatric culture as well as experimental and artists' film. Recent publications include a co-edited volume of the film studies journal *Décadrages* on Expanded Cinema (forthcoming) and an essay in *Marcel Duchamp and the Forestay Waterfall*, ed, Stefan Banz (2010).

Catherine Mills is a Senior Lecturer in Bioethics at Monash University. She is the author of two books, *Futures of Reproduction: Bioethics and Biopolitics* (2011) and *The Philosophy of Agamben* (2008), as well as numerous articles in political theory, feminist theory and bioethics. Her current research focuses on issues in biopolitics and bioethics, especially pertaining to human reproduction.

Warren Neidich is an artist and theorist who works between Los Angeles and Berlin. His work has focused on how artistic interventions create places for the disorganization and then reorganization of the understanding, in an inversion of Joseph Kosuth's famous formula: "Art before Philosophy, not After." Recent awards include The Murray and Vickie Pepper Distinguished Visiting Artist and Scholar Award, Pitzer College, 2012, The Fulbright Scholar Program Fellowship, Fine Arts Category, 2011 and the Vilem Flusser Theory Award, Berlin, 2010.

Jakob Nilsson received his PhD from the Department of Cinema Studies, Stockholm University, and the Research School of Aesthetics, Stockholm University. His doctoral thesis, *The Untimely-Image: On Contours of the New in Political Film-Thinking*, was published in 2012. He has contributed articles for *Journal of Aesthetics and Culture*, *SITE*, and *Rhizomes*.

Johanna Oksala is Academy Research Fellow in the Department of Philosophy, History, Culture and Art Studies at the University of Helsinki. She is the author of *Foucault on Freedom* (2005), *How to Read Foucault* (2007), *Foucault, Politics, and Violence* (2012) and *Political Philosophy* (forthcoming in 2013).

Julian Reid has taught at the School of Oriental and African Studies, University of London, Sussex University, King's College London, and the University of Lapland (Finland) where he currently holds the Chair in International Relations. His publications include *The Biopolitics of the War on Terror* (2006) and *The Liberal Way of War* (with Michael Dillon). His new book, *Deleuze & Fascism* (co-edited with Brad Evans) is forthcoming in 2013.

Cecilia Sjöholm is Professor of Aesthetics at Södertörn University. Her books include *Aisthesis, estetikens historia del 1* (ed. with Sara Danius and Sven-Olov Wallenstein, 2012), *Translatability* (ed. with Sara Arrhenius and Magnus Bergh, 2011), *Kristeva and the Political* (2005), *The Antigone Complex; Ethics and the Invention of Feminine Desire* (2004), *Föreställningar om det omedvetna: Stagnelius, Ekelöf and Norén* (1996), and *Ensam och pervers* (with Sara Arrhenius, 1995.) Her research interests are currently focused on issues in phenomenology and aesthetics, with a particular view on the history of aesthetics and its relation to politics.

Łukasz Stanek has taught at the ETH Zurich and Harvard GSD, and is the 2011-13 A. W. Mellon Post-Doctoral Fellow at the Center for Advanced Study in the Visual Arts (CASVA) at the National Gallery of Art in Washington, D.C. Stanek authored *Henri Lefebvre on Space: Architecture, Urban Research, and the Production of Theory* (2011) and he is currently editing Lefebvre's unpublished book on architecture. Stanek's second field of research is the export of architecture and urbanism from socialist countries to Africa, Asia, and the Middle East during the Cold War. On this topic, he co-edited a recent issue of *The Journal of Architecture* (17:3, 2012).

Sven-Olov Wallenstein is Professor of Philosophy at Södertörn University in Stockholm, and is the editor-in-chief of *Site*. He is the translator of works by Baumgarten, Winckelmann, Lessing, Kant, Hegel, Frege, Husserl, Heidegger, Levinas, Foucault, Derrida, Deleuze, and Agamben, as well as the author of numerous books on contemporary philosophy, art, and architecture. Recent publications include *Essays, Lectures* (2007); *Thinking Worlds: The Moscow Conference on Art, Philosophy and Politics* (co-ed. with Daniel Birnbaum and Joseph Backstein, 2007); *The Silences of Mies* (2008); *Biopolitics and the Emergence of Modern Architecture* (2009); *1930/31: Swedish Modernism at the Crossroads* (with Helena Mattsson, 2009); *Swedish Modernism: Architecture, Consumption and the Welfare State* (co-ed. with Helena Mattsson, 2010), *Nihilism, Art, Technology* (2011), and *Translating*

Hegel: The Phenomenology of Spirit and Modern Philosophy (co-ed. with Brian Manning Delaney, 2012).

Södertörn Philosophical Studies

Södertörn Philosophical Studies is a book series published under the direction of the Department of Philosophy at Södertörn University. The series consists of monographs and anthologies in philosophy, with a special focus on the Continental-European tradition. It seeks to provide a platform for innovative contemporary philosophical research. The volumes are published mainly in English and Swedish. The series is edited by Marcia Sá Cavalcante Schuback and Hans Ruin.